# Springer Theses

Recognizing Outstanding Ph.D. Research

## Aims and Scope

The series "Springer Theses" brings together a selection of the very best Ph.D. theses from around the world and across the physical sciences. Nominated and endorsed by two recognized specialists, each published volume has been selected for its scientific excellence and the high impact of its contents for the pertinent field of research. For greater accessibility to non-specialists, the published versions include an extended introduction, as well as a foreword by the student's supervisor explaining the special relevance of the work for the field. As a whole, the series will provide a valuable resource both for newcomers to the research fields described, and for other scientists seeking detailed background information on special questions. Finally, it provides an accredited documentation of the valuable contributions made by today's younger generation of scientists.

## Theses are accepted into the series by invited nomination only and must fulfill all of the following criteria

- They must be written in good English.
- The topic should fall within the confines of Chemistry, Physics, Earth Sciences, Engineering and related interdisciplinary fields such as Materials, Nanoscience, Chemical Engineering, Complex Systems and Biophysics.
- The work reported in the thesis must represent a significant scientific advance.
- If the thesis includes previously published material, permission to reproduce this must be gained from the respective copyright holder.
- They must have been examined and passed during the 12 months prior to nomination.
- Each thesis should include a foreword by the supervisor outlining the significance of its content.
- The theses should have a clearly defined structure including an introduction accessible to scientists not expert in that particular field.

More information about this series at http://www.springer.com/series/8790

Yanlan Liu

# Multifunctional Nanoprobes

From Design Validation to Biomedical
Applications

Doctoral thesis accepted by Changchun Institute of Applied
Chemistry, Chinese Academy of Sciences, Changchun,
China

 Springer

*Author*
Dr. Yanlan Liu
State Key Laboratory of Electroanalytical
 Chemistry, Changchun Institute of
 Applied Chemistry
Chinese Academy of Sciences
Changchun
China

and

Brigham and Women's Hospital
Harvard Medical School
Boston, MA
USA

*Supervisor*
Prof. Lehui Lu
State Key Laboratory of Electroanalytical
 Chemistry, Changchun Institute of
 Applied Chemistry
Chinese Academy of Sciences
Changchun
China

ISSN 2190-5053          ISSN 2190-5061   (electronic)
Springer Theses
ISBN 978-981-13-5586-8        ISBN 978-981-10-6168-4   (eBook)
https://doi.org/10.1007/978-981-10-6168-4

Printed on acid-free paper

This Springer imprint is published by Springer Nature
The registered company is Springer Nature Singapore Pte Ltd.
The registered company address is: 152 Beach Road, #21-01/04 Gateway East, Singapore 189721, Singapore

# Supervisor Foreword

Molecular imaging technology is the cornerstone of various analytical applications, in particular for the biomedical field. Traditional molecular imaging probes are based on small molecules, which have shown great limitations regarding the specificity and imaging effect. Recent advances in the development of nanotechnology and materials science have demonstrated tremendous opportunities in biomedicine, given the unique phy-chemical properties of nanomaterials including controllable synthesis, easy surface modification, and enhanced accumulation in the area of interest. In recent years, nanoprobes has been emerging as the hot topic of research and also playing an increasingly important role in various biomedical applications such as the detection of toxic substances, diagnosis, and analysis of human diseases both in vitro and in vivo.

Multifunctional imaging nanoprobes that combine different imaging modalities can compensate for the deficiencies of individual imaging modalities, thus providing solid and more accurate information for biomedical analysis. Moreover, the incorporation of different imaging and therapeutic functions within a single nanoformulation, which is referred to as nanotheranostics, is of high interest and significance for real-time tracking of the pharmacokinetics and biodistribution of the therapeutic drugs, patient selection, and personalized treatments. Although numerous efforts have been made, currently studied multifunctional nanoprobes are far from satisfied for potential clinical translation.

This thesis integrates the knowledges from chemistry, material science, and biomedicine, and reports the design and synthesis of innovative multifunctional nanoprobes. It includes novel methods for the preparation of multifunctional nanoprobes and important theoretical fundamentals in the design of nanoprobes, aiming at addressing the problems associated with currently available multifunctional nanoprobes including clinically used imaging contrast agents, in terms of the synthesis, imaging performance, and biocompatibility. I would expect that this thesis would be beneficial for guiding future studies in the development of novel multifunctional nanomaterials with significantly enhanced performance.

Changchun, China                                        Lehui Lu

# Abstract

Nanomaterials have shown high superiority over bulk materials and small molecules in the applications of biomedicine, due to many advantages such as controllable synthesis, easy modification, long-term circulation in vivo, facile integration of multifunctionalities into a single nanoformulation, and passive or negative accumulation in the targeted sites. Therefore, nanomaterials have found broad applications in many fields of biomedicine and have been attracting increasing attention. To address the issues of nanoprobes regarding the synthesis and performance in the biomedical fields, this dissertation is mainly focused on the design and synthesis of novel multifunctional nanoprobes depending on different requirements, and investigated their potential use in the applications of biosensing, molecular imaging, and theranostic treatment of cancer. The main points are summarized as following:

Chapter 1. In this chapter, we have provided a summary regarding the concept, properties, and biomedical applications of nanomaterials.

Chapter 2. Cyanide is a highly toxic substance and can invade the human body through many routes. Nevertheless, cyanide is still used in many practical fields. Thus, there is an urgent need to develop highly sensitive and selective sensing systems for the detection of cyanide in the environment, especially in water and biological samples. To overcome the problems of currently studied cyanide sensors such as poor water solubility, poor selectivity, and complex preparation procedures, we develop an innovative gold-nanocluster-based fluorescent sensor for cyanide in aqueous solutions. Owing to the unique Elsner reaction between cyanide and the gold atoms of gold nanoclusters, this sensor shows high sensitivity and strong tolerance to other interferences. More impressively, this sensor can be directly used for the detection of cyanide in the aqueous solution with excellent recoveries. Thus, this gold-nanocluster-based sensor may provide an effective new tool for highly sensitive and selective detection of cyanide in biological samples.

Chapter 3. Although the fluorescence imaging (FI) technique has high sensitivity, its penetration capability is very limited. In contrast, magnetic resonance imaging (MRI) has deep tissue penetration but low sensitivity. Thus, more available and accurate diagnostic information can be anticipated after combination of FI with MRI. However, currently existent methods for the fabrication of MRI-FI nanoprobes are complicated, and the resultant MRI-FI nanoprobes have demonstrated many disadvantages such as high toxicity, large particle size, or low relaxivity and quantum yield. To address these issues, we described in this chapter a straightforward and versatile method to develop MRI-FI dual modality nanoprobes by doping $Gd^{3+}$ ions in low toxic ZnO quantum dots (QDs). The resultant Gd-doped ZnO QDs are ultrasmall in size and have enhanced fluorescence resulting from the Gd doping. In vitro experiments confirm that Gd-doped ZnO QDs can successfully label the HeLa cells in short time and present no evidence of toxicity or adverse effects on cell growth. Besides, they exert a strong positive contrast effect with a large longitudinal relaxivity much higher than that of Gd-based clinical MRI contrast agent.

Chapter 4. X-ray computed tomography (CT) has been considered to be the most powerful diagnostic tool in clinical diagnosis due to its many advantages compared to other molecular imaging tools. X-ray CT contrast agents currently used in clinic scanning are mainly based on iodinated small molecules. However, these small molecules suffer from many disadvantages, such as low contrast efficiency, very short circulation lifetime, and potential renal toxicity. Moreover, some patients are hypersensitive to iodine. These disadvantages have significantly restricted the applications of X-ray CT in biomedicine, particularly in targeted imaging. In this chapter, we describe a first-in-class Yb-based nanoparticulate CT contrast agent. Owing to the attenuation characteristics of Yb, which is matched with the X-ray photon energy used in clinical applications, the Yb-based nanoparticulate CT contrast agent offers a much higher contrast efficacy compared to the clinical iodinated agent at 120 kVp. Along with long circulation time and low toxicity in vivo, these nanoparticles can act as a high-performance CT contrast agent for in vivo angiography and bimodal image-guided lymph node mapping. By doping Gd into the nanoparticles, this contrast agent also shows enhanced upconversion luminescence and MRI capability.

Chapter 5. In clinical CT examination, the operating voltage changes from 80 to 140 KVp depending on the applications. All the contrast agents that contain single contrast element can only provide limited contrast efficacy and cannot be tailored to the changes of the operating voltage. To overcome this hurdle, we describe, in this chapter, a binary nanoparticulate CT contrast agent by integrating Yb and Ba into a single nanoparticle. Owing to the big difference between the K-edge energy of these elements, this binary contrast agent can provide much higher contrast efficiency than that of the clinical iodinated contrast agent, regardless of the voltage setting. More importantly, this agent has low toxicity and further improved in vivo imaging effect, which is successfully used for in vivo X-ray computed tomography angiography.

Chapter 6. Owing to high selectivity and minimal invasiveness, photothermal therapy is emerging as a powerful technique for cancer treatment. However, currently available photothermal therapeutic (PTT) agents have not yet achieved clinical implementation, stemming from great concerns regarding their long-term safety. From this point of view, we develop in this chapter a novel PTT agent based on dopamine-melanin colloidal nanospheres. Benefiting from wide distribution of their component in human naturally, this new PTT agent shows biodegradability, a high median lethal dose, and does not induce long-term toxicity during their retention in rats. Moreover, this agent can offer much higher photothermal conversion efficiency than previously reported PTT agents. Upon irradiation with 808 nm laser, dopamine-melanin colloidal nanospheres can efficiently absorb light and transfer it into heat. Both in vitro and in vivo experiments prove that these nanospheres can destroy tumor tissue and inhibit the regrowth of the tumor. Furthermore, dopamine-melanin colloidal nanospheres can be easily attached to conjugates with other interesting biofunctionalities. By covalent modification of Gd-DTPA, MRI-guided tumor targeted photothermal therapy is achieved.

**Keywords** Nanomaterials · Probes · Molecular imaging · Sensing · Multifunction Cancer therapy

# Acknowledgements

Financial support by NSFC (No. 21125521, 21075117), the National Basic Research Program of China (973 Program, No. 2010CB933600) and the "Hundred Talents Project" of the Chinese Academy of Science, and NSFC (No. 20873138) is gratefully acknowledged.

# Contents

# Chapter 1
# Literature Review

**Abstract** In this chapter, we have provided a summary regarding the concept, properties, and biomedical applications of nanomaterials.

## 1.1 Introduction of Nanomaterials

Nanotechnology is a new field of science that established on nowadays advanced scientific technology. The idea and hypothesis of nanotechnology was first suggested by a renowned American physicist Richard P. Feynman in 1959 in one of his report. As a pioneer of nanotechnology, he predicted that humans would be able to control the assembly of molecules and atoms using their will. The rapid development of scientific technology in the 1980s has led to the invention of AFM, STM, calculator and TEM, which allows one to examine materials on a nanoscale. These technologies promote the application of nanotechnology in a variety of major fields of science, such as physics, chemistry, biomedicine, material, energy, and electronics. With the development of advanced technologies, nanotechnology has been integrated into different aspects of human life, generating a profound impact to human lifestyle. Scientists admit that the development of nanotechnology gives an enormous boost to the economic growth in the 21st century; its impact on world development is significantly greater than that of the development of microelectronics in the late 20th century. The development of nanotechnology brings human history to a new era, and it will be a major theme of the next technological revolution.

Nanomaterials are the core elements of nanotechnology. Generally, nanomaterials are referring to materials that have at least one dimension sized between 1–100 nm in a three dimension setting, or complex materials that are assembled by these materials. Due to their small size, nanomaterials display many unique physical and chemical properties when compared to bulk materials. The following lists the general properties of nanomaterials:

© Springer Nature Singapore Pte Ltd. 2018
Y. Liu, *Multifunctional Nanoprobes*, Springer Theses,
DOI 10.1007/978-981-10-6168-4_1

1. Small size effect

Small size effect refers to the change in macroscopic physical and chemical properties that occurs when the size of nanomaterials is reduced to a certain value. This effect is well demonstrated by gold and silver nanoparticles. When the size or the surface properties of the gold or silver nanoparticles changes, the surface plasmon resonance (SPR) of the nanoparticles also changes, followed by the change in color of the solution [1, 2].

2. Surface effect

When the size of nanomaterials decreases, the number of atoms on the material surface and the total number of atoms increase significantly, leading to a remarkable change in the surface area and surface energy. This phenomenon is denoted as the surface effect. Taking advantage of this effect, researchers can increase the surface area of the nanomaterial by changing its size in order to enhance some specific properties of the material.

3. Quantum size effect

The electron energy level near the Fermi level of bulk metal materials can be seen as continuous. When the dimension of the material decreases to nanoscale, its energy band will split into discrete energy levels, and this phenomenon is known as the quantization of energy level. The gap between energy levels increases with the decrease of dimension. When the change in energy level is greater than the change in thermal, photonic, electromagnetic, or superconducting condensation energy, the material will exhibit many properties that are much different from regular materials. This effect is called the quantum size effect.

Apart from this, nanomaterials also process many other properties, such as bulk effect, macroscopic quantum tunneling effect, dielectric confinement effect, and so on. Nanomaterials have been widely used in different areas because of their unique photonic, electronic, magnetic and chemical properties when compared to bulk materials. This chapter focuses on the applications of nanomaterials in the biomedicine.

## 1.2  Advantages of Nanoprobes in Biomedical Applications

Industrial development has provided a lot of benefits to the society, yet, it also increases the threats to human health. Improper waste disposal and leakage of toxic materials have resulted in severe consequences. It is clear that the incidence of cancer increases gradually per year. It is roughly estimated that there are about 1.4 to 1.5 million people that have died of cancer in China annually. While cancer mortality has doubled in the past 30 years, only about 0.7 million people died of cancer in the 1970s [3]. In addition to cancer, the incidence of other diseases such as cardiovascular disease and diabetes also increases constantly, and the patients tend to get younger. However, limitations in modern clinical diagnostic and therapeutic techniques make it more difficult to treat these diseases.

Biomedical science has been deemed to be one of the most important fields of science in the 21st century. Nanomedicine, which refers to the application of nanomaterials in biomedicine, has been bought to the forefront of scientific research worldwide. Apart from the properties mentioned above, there are other features of nanomaterials that lead to its high popularity among researchers:

1. Controllable synthesis

The ability to control the synthesis of nanomaterials is a major factor that makes nanomaterials important in biomedicine. With the advanced technological development in the synthesis of nanomaterials, the size and composition of nanomaterials can be adjusted. Nanomaterials that vary in size and morphologies have been widely studied. From a biomedical point of view, the in vivo distribution and metabolic rate of nanomaterials are significantly affected by their size and morphology. For instance, studies have shown that cells react differently when encountering silica in different morphologies. Large rod-shaped silica is internalized by cells more quickly and in a higher quantity than small sphere-shaped silica. Meanwhile, the cell adhesion, viability, and differentiation depend largely on the morphology and size of the nanoparticles [4–7].

Adjusting the composition of nanomaterials always creates new features such as fluorescent nanomaterials, magnetic nanomaterials, surface plasmon nanomaterials, etc. [8–10]. Each type of materials can be classified into different categories, where fluorescence nanomaterials are the most typical example. Traditionally, fluorescent nanomaterials are primarily based on transition metal doped semiconductor quantum dots such as CdSe, CdS, etc. By adjusting their compositions, fluorescent nanomaterials can emit fluorescence with different wavelengths [11]. Since these materials are highly toxic, scientists have recently invented upconversion fluorescent nanomaterials and fluorescent gold nanoclusters [12]. Compared to organic fluorescent molecules, fluorescent nanomaterials have higher photostability and hence are less prone to photobleaching. Thus, fluorescent nanomaterials are expected to be promising candidates for fluorescence imaging analysis.

2. Ease of functional surface modification

Another important feature of nanomaterials is that it is easy to modify their surface with functionalities. Compared to small molecule materials, various molecules such as proteins, targeted molecules, and molecules with other functions can be conjugated to the surface of nanomaterials via covalent or non-covalent interactions. Functionalized nanomaterials can be obtained easily through centrifugation and washing, without the need of complex distillation or organic synthesis. The biggest concern of the currently available imaging contrast agents and cancer therapeutic drugs for clinical diagnosis is their specificity. This problem can be overcome by the use of nanomaterials, with which targeted molecular imaging, targeted drug delivery, and targeted labeling, etc. are possible. To conclude, surface-functionalized nanoparticles can be classified into two major categories:

2.1. Covalent interaction. The synthesis of hydrophilic nanomaterials always involves the use of certain protective agents and stabilizers, or the nano-materials themselves contain hydroxyl, carboxyl, amino, epoxy, or other groups on their surface. The presence of these groups allows for func-tionalization of the nanomaterials via covalent modifications. The most typical nanomaterial is the silica nanoparticles, in which the oxygen and the silicon atoms are connected together via the covalent bond. Thus, the surface of silica nanoparticles contains a large number of hydroxyl groups. Through hydrolysis, the surface of silica nanoparticles can be modified by silioxane molecules with different functional groups. Currently, a lot of studies used silioxane molecules with amino, carboxyl, epoxy, or azide groups to modify the surface of silica nanoparticles, which facilitated the conjugation of targeting ligand to the silica nanoparticles via covalent interactions. Due to its high biocompatibility, silica nanoparticles have shown great potential use in drug delivery and molecular imaging [13–15].

2.2. Non-covalent interaction. Surface modification of nanomaterials via the non-covalent interaction can be classified into three major categories. In order to effectively control the morphology, dimension, and dispersity of nanomaterials, researchers often choose hydrophobic ligands as a protec-tive agent and stabilizer to synthesize different nanomaterials. These resulting nanomaterials are generally hydrophobic, and they have to be converted to hydrophilic nanomaterials prior to biomedical applications. One way to achieve this is to perform a ligand exchange with a suitable hydrophilic ligand. The choice of hydrophilic ligand depends on the composition of the nanomaterials. For instance, Sandiford et al. prepared oleylamine-coated ultrasmall-superparamagnetic oxide nanomaterials, phosphate-containing polyethylene glycol (PEG) was thus selected for ligand exchange in order to modify the surface of the iron oxide nanoparticles, given the strong coordination between phosphate and iron ions. The PEGylated iron oxide nanoparticles have been successfully used in magnetic resonance imaging in vivo [16]. Similarly, Gao et al. used asymmetric PEGs with maleimide at one end and phosphate at the other to modify the surface of fluorescent nanomaterials, and high monodispersity of the nanoparticles is retained. Meanwhile, antibody has been added to the surface of the nanomaterials through the "click" reaction with maleimide, which allows for targeted tumor imaging. In addition to PEG, commonly used reagents for ligand exchange include polyvinyl pyrrolidone (PVP), mercaptopropionic acid, polyacrylic acid, etc. [17–21]

The second method is based on hydrophobic-hydrophobic interaction. Amphiphilic molecules are often chosen as the ligand in this modification process. The hydrophobic end of the amphiphilic molecule forms hydrophobic -hydrophobic interaction with the hydrophobic surface of the nanomaterial, the hydrophilic end of the amphiphilic molecule is now pointing outwards and hence the nanomaterial is converted from hydrophobic to hydrophilic. For example: In 2012, Gu et al. used the compound formed by reacting folic acid conjugated amiphiphilic chitosan with long chain alkane as a ligand for surface modification of the fluorescent nanomaterial, then porphyrin was embedded in the nanomaterial via the hydrophobic - hydrophobic interaction, which allows for targeted imaging and tumor therapy, shown in Fig. 1.1 [22].

In addition to the above two methods, silica coating is undoubtedly another effective method for surface modification of nanomaterials. Different from the above two methods, silica coating can be performed for both hydrophilic and hydrophobic materials, and the thickness of the silica coating can be well controlled through adjusting the amount of precursor. Typically, either hydrophobic or hydrophilic silica-coated nanomaterials are fabricated by microemulsion. The resulting silica shell not only changes the surface property of the nanomaterials, but also allows further functional modification and reduces the toxicity of the nano-materials, etc. Most recently, researchers have applied this method to a variety of nanomaterials, including gold nanostructures, magnetic nanomaterials, upconver-sion fluorescent nanomaterials, and semiconductor quantum dots, etc. [23–27]. Furthermore, introducing surfactants in the process of silica coating can result in a

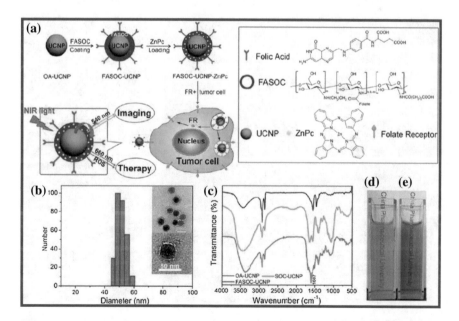

**Fig. 1.1** Surface modification of nanomaterials through the hydrophobic interaction [22]. (Copyright 2013 American Chemical Society)

mesoporous silica shell, where drugs or other materials can be loaded in the pores. For example: In 2008, Kim et al. utilized this method to coat a layer of uniform mesoporous silica shell on iron oxide nanoparticles, as shown in Fig. 1.2. After loading of fluorescent dyes and drugs, fluorescence imaging, magnetic resonance imaging and drug release could be achieved [24].

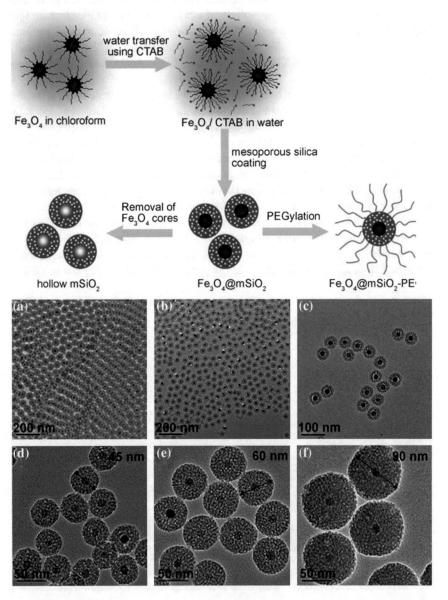

**Fig. 1.2** The synthesis of mesoporous silica-coated $Fe_3O_4$ nanoparticles [24]. (Copyright 2008 Wiley-VCH)

3. Loading of various multifunctional materials

Another remarkable feature of nanomaterials is that different functionalities can be integrated into a single nanoparticle. The versatility of nanomaterials has led to its wide application in the biomedical field. The multi-functionality of nanoparticles can be achieved by various means, such as doping, coating, covalent linkage, etc. [28–30]

4. Long circulation time in vivo

Apart from the aforementioned physical and chemical properties, the advantages of nanoscale size itself cannot be ignored, as it makes nanomaterials ideal for biomedical applications. The kidney filtration system allows a maximum particle size of $\sim 10$ nm to pass through. Currently, drugs and imaging agents in clinical use are mainly composed of small molecular species. Due to the small molecular weight, small molecules will be excreted by kidney quickly upon intravenous injection. In contrast, a large number of studies have shown that nanomaterials with the size between 10–200 nm and appropriate surface modification have long in vivo circulation time. Meanwhile, nanomaterials can be effectively accumulated at the tumor site through the enhanced permeability and retention (EPR) effect [31–35].

## 1.3 Application of Nanoprobes in Life Analysis

Nanomaterials are in every aspect of life science, and their advantages in biomedical applications have been well demonstrated. In this section, the application of nanoprobes in biosensing, molecular imaging, and cancer treatment will be introduced.

### 1.3.1 Application of Nanoprobes in Biosensing

Since nanomaterials have unique optical, electronic, and magnetic properties, their application in environmental toxicology for the detection of toxic and harmful materials has received increasing attention. Biologically, the introduction of nanomaterials and their combination with various spectroscopic techniques greatly improve the performance of the biosensors. Compared to conventional biological detection techniques such as ELISA, nano-bioprobes have numerous advantages including: (1) low cost; (2) easy to operate, as the use of nano-bioprobes usually does not require large and expensive instruments, and the training for analyst is simple; (3) robust and facile synthesis; (4) diversity and universality, as a large variety of nanomaterials is available, and many substances could be detected simultaneously by selecting suitable nanomaterials and ligands; (5) high sensitivity, selectivity, and rapid detection capability of functionalized nanomaterials [36].

Depending on the functions of the nanomaterials, nanobiosensors are different in detection methods, the following lists the common detection methods:

1. Fluorescence analysis

Fluorescence analysis has the highest sensitivity among all the detection methods. Therefore, this method is the most suitable for the detection of trace substances. In addition, fluorescence analysis is simple and has fast response time. Currently, fluorescence analysis is a popular detection method in biosensing and has been widely used for detecting protein, organic small molecules, DNA, and other toxic, harmful substances in the human body [37–40]. According to the response mechanism, fluorescent sensors can be classified into three major categories: turn-on, turn-off, and ratiometric.

Turn-off fluorescence sensor is the most typical fluorescence sensing method. Its detection mechanism is based on direct or indirect quenches of the fluorescence of the sensor by the reaction between the analyte and the sensor/surface ligands. The use of turn-off fluorescence sensors has been reported in the detection of various biomoleculars [41–45]. A typical example has been demonstrated by Tang et al. [45]. Tang et al. modified the surface of CdTe semiconductor quantum dots with glucose oxidase to create a glucose sensor. When glucose is presented in the analyte, the glucose oxidase on the surface of the quantum dots would catalyze the oxidation of glucose into hydrogen peroxide, and the hydrogen peroxide would quench the fluorescence of the quantum dots. With this method, glucose can be detected in the cellular level, as shown in Fig. 1.3.

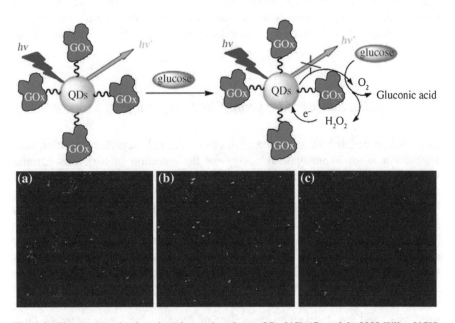

**Fig. 1.3** Glucose detection based on the semiconductor QDs [45]. (Copyright 2008 Wiley-VCH)

The detection mechanism of turn-on fluorescence sensors is based on the certain reactions between the analyte and the sensor itself or its surface ligands, resulting in fluorescence recovery of the sensor or emission of fluorescence from non-fluorescent materials. Compared with turn-off fluorescence sensors, turn-on fluorescence sensors have lower background signals, thus the result is more accurate and reliable. A large number of studies on the turn-on fluorescence sensor have been published [46–50]. For example, Liu et al. reported a glutathione detection system using upconversion fluorescent nanoparticles in 2011 (Fig. 1.4). The upconversion fluorescent nanoparticles emit blue fluorescence upon 980 nm laser irradiation, where the fluorescence would be quenched when the surface of the nanoparticles is covered by $MnO_2$. Nonetheless, when glutathione is present in the analyte, glutathione would convert the $MnO_2$ molecules on the nanoparticles' surface to $Mn^{2+}$ ions, and thus allows the recovery of the fluorescence. This probe has an outstanding sensitivity with the lowest detection limit is 0.9 μM, and the detection of glutathione is not interfered by other substance. This probe has been proven to be effective for intracellular glutathione detection [46].

Apart from small biological molecules, turn-on fluorescence sensor is also commonly used in large molecules such as proteins, pathogens, and even cells detection. Recently, Rotello et al. developed a "chemical nose" sensor for the

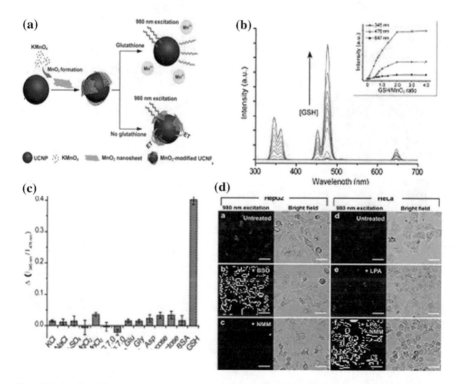

**Fig. 1.4** A glutathione detection system using upconversion fluorescent nanoparticles [46]. (Copyright 2011 American Chemical Society)

detection and identification of proteins. As shown in Fig. 1.5, the author modified the surface of the gold nanoparticles with different cationic ligands, and then coated a layer of anionic fluorescent polymer. Since gold nanoparticles have strong fluorescence quenching capability, the fluorescence of the polymer on the nanoparticles' surface is quenched. However, when the targeted proteins are present in the solution,

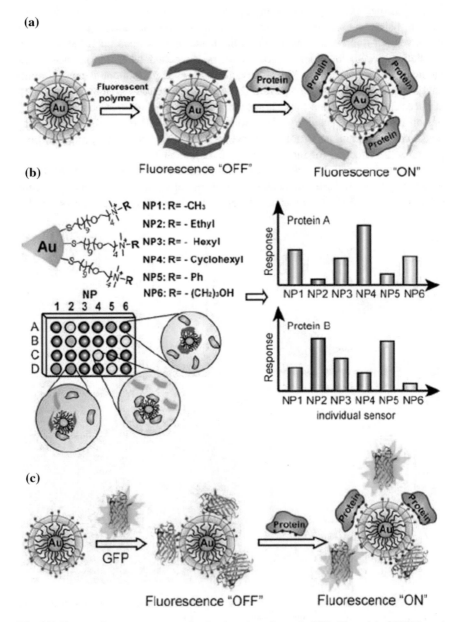

**Fig. 1.5** Turn-on fluorescence sensor for the detection of protein [51]. (Copyright 2007 Nature)

the proteins would compete with the fluorescent polymer for the active sites on the surface of the gold nanoparticles, causing the fluorescent polymer to detach from the surface of gold nanoparticles and subsequent fluorescence recovery of the polymer. As different types of ligands have been added to the surface of the gold nanoparticles, and the coordination ability is different between different proteins and the gold nanoparticles, the detection of specific proteins is possible.

Ratiometric fluorescent probes generally emit two light beams of different wavelengths upon excitation. The presence of analyte would increase or decrease the intensity of one of the light beam while the intensity of the other one remains unchanged, or increase the intensity of one of the light beam as the intensity of the other one decrease. Base on this detection principle, ratiometric fluorescent probes have higher sensitivity than the other two types fluorescent probes mentioned above, and hence its application in biosensing is being developed in recent years [52–55].

2. Colorimetry

Noble metal nanoparticles such as Au and Ag possess surface plasmon resonance (SPR) properties, and SPR is closely related to the dimension of the nanoparticles. When gold and silver nanoparticles aggregate, the surface plasmon absorption of the nanoparticles will change, leading to a color change in the solvent of the nanoparticles. Taking advantage of this property, gold and silver nanoparticles are widely used for colorimetric analysis [56–60]. This method is very convenient as the color change can be observed visually. Furthermore, the detection technique only involves ultraviolet light absorption of the detection solvent, thus this detection method is easy and cheap to operate. Our group has applied this method in the past and we reported for the first time using cyanuric acid-modified gold nanoparticles for visual detection of melamine successfully, as demonstrated in Fig. 1.6 [61].

Nowadays, the incidence of diabetes is increasing, and the proportion of young patients is constantly rising. Unfortunately, there is no effective treatment for diabetes up to date. Diabetes can be effectively avoided by monitoring the blood glucose level. Recently, colorimetric detection of blood glucose has been reported using gold nanoparticles [62–65]. For example, Aslan et al. prepared dextran-coated gold colloids, and the addition of Con A protein to the gold colloids result in formation of cross-linkages between Con A protein and dextran on the gold colloids, which causes gold colloids aggregation. When glucose is present in the analyte, the dextran-coated gold colloids would disperse again since the cross-linking of Con A with glucose is stronger than the cross-linking of Con A with dextran, and hence the color of the solvent changes from blue to red. This system detects glucose with a detection limit of 1–40 mM, and can be directly applied to test the amount of glucose in the urine samples of diabetic patients (Fig. 1.7) [62].

3. Electrochemical detection

In addition to fluorescent and colorimetric methods, electrochemical detection is another commonly used method for the detection of biological molecules. The advantages of

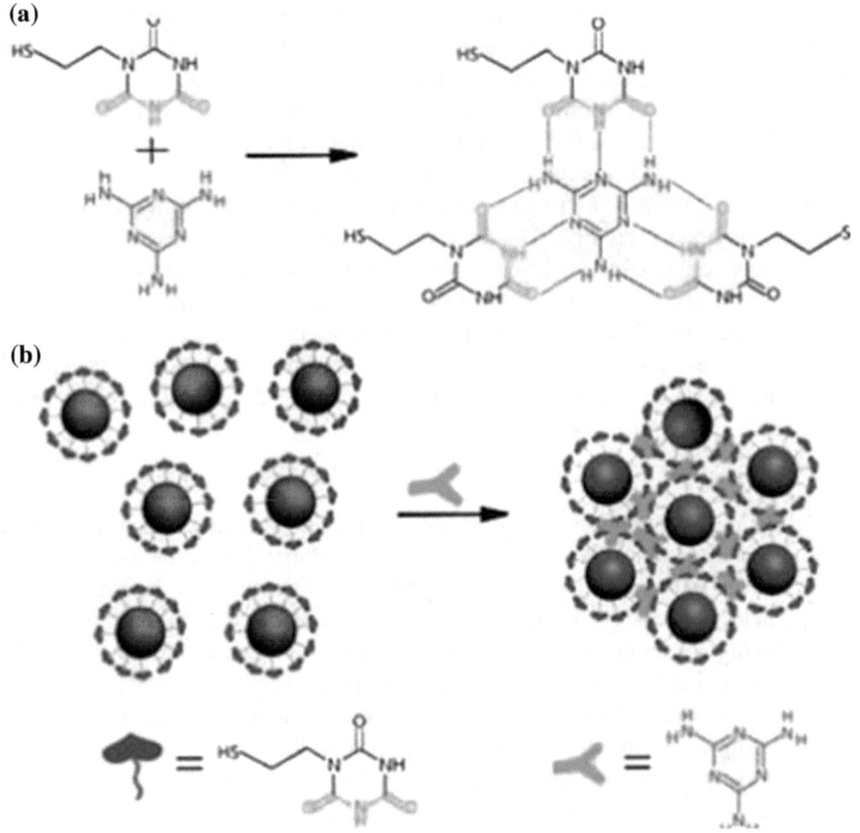

**(a)**

**(b)**

**Fig. 1.6** Colorimetric detection of melamine using Au nanoparticles [61]. (Copyright 2009 American Chemical Society)

electrochemical detection include rapid and simple procedures to perform, high selectivity and sensitivity, and broad concentration detection range etc., which make electrochemical detection very attractive in biomedical science. Research on electrochemical biosensors mostly focuses on using biological molecules, enzymes, or antibodies as the recognition unit to construct highly selective and sensitive electrochemical biosensor systems [66–70]. Gao et al. constructed an electrochemical sensor system for glucose by grafting glucose oxidase onto the carbon nanotube/Ag electrode material. This electrochemical sensor is very effective in glucose detection with a detection limit of 17 μM. In order to detect samples in trace amount, signal amplification strategy is frequently used in electrochemical sensing. For example, with a signal amplification super sandwich strategy, Liu et al. developed an electrochemical sensing platform for DNA detection [66]. With polydopamine as the linker, gold nanoparticles were coated onto the carbon nanotubes. The selectivity of DNA and cancer cells can be achieved through selecting different DNA aptamers, as shown in Fig. 1.8.

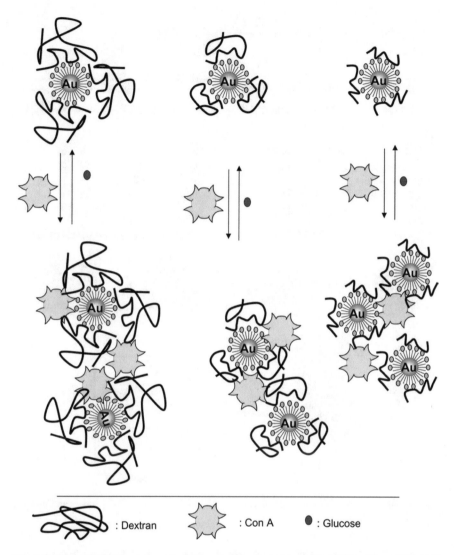

**Fig. 1.7** Colorimetric detection of glucose using Au nanoparticles [62]. (Copyright 2004 American Chemical Society)

### 1.3.2  Application of Nanoprobes in Molecular Imaging

In 1895, the German scientist Röntgen discovered X-rays and successfully used it to take a picture of his wife's hand, which leads to a revolution of medical diagnosis [71]. In 1969, British scientist Hounsfield developed X-ray computed tomography by combining the characteristics of calculator and X-ray. Thereafter, the development of medical imaging sharply increases since the invention of magnetic

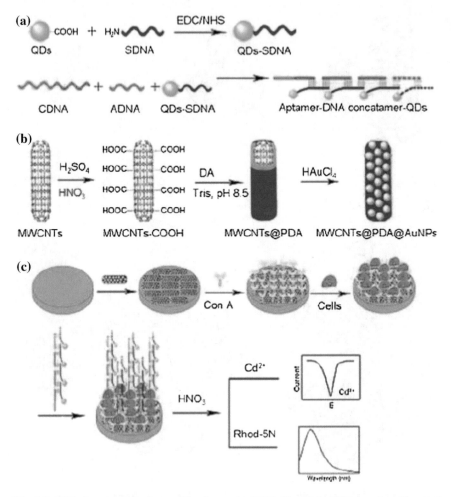

**Fig. 1.8** DNA detection with the amplification method [66]. (Copyright 2013 American Chemical Society)

resonance imaging in 1973. Along with the techniques of conventional imaging systems, scientists combine molecular biology, chemistry, physics, and computer science to form a new discipline, which is denoted as molecular imaging (MI) [72]. The concept of molecular imaging was first proposed by Dr. Weissleder at the Harvard University in 1999 [73]. It relies on particular molecules in the body as contrast agents and uses existing molecular imaging techniques as technical support for non-invasive real-time imaging analysis of the human body to obtain human anatomical information through subsequent computational analysis, which allows observation of changes at molecular level and cellular level in vivo to understand the pathology of diseases [74]. With advancing technological development in nanotechnology and electronic science, novel molecular imaging techniques have been developed. The role of molecular imaging techniques involving the use of nanomaterials is getting more significant in early medical diagnosis, drug development, and new diagnostic technology.

### 1.3.2.1 Application of Nanoprobes in Optical Imaging

The most common optical imaging technique in biomedical science is fluorescence imaging. Compared with other imaging systems, fluorescence imaging has the following advantages: (1) high sensitivity and cellular-level detection; (2) easy operation at low cost; and (3) safe to work with as no ionizing radiation is involved and hence there are no health threats to long-term workers. Fluorescence imaging has thus been widely applied in various biomedical researches, including in vivo tracking of tumor growth, distribution, and metastasis, etc. [75–80]. The most typical fluorescent nanomaterials are semiconductor quantum dots. In 1998, Alivisatos et al. and Nie et al. published in Science about using quantum dots as bioprobes, they reported for the first time using quantum dots as the fluorescent biomarker for labeling of the living cell system, which has raise scientists' interest in quantum dots research worldwide [81, 82]. The biggest advantage of semiconductor quantum dots is adjustable emission spectrum (Fig. 1.9). Therefore, semiconductor quantum dots have been extensively studied in biological labeling. Nonetheless, this type of semiconductor quantum dots is composed of heavy metals such as Cd, which increase the toxicity of the quantum dots. To address this safety issue, researchers have been seeking for ways to reduce the toxicity of the quantum dots. The most typical method is to coat a layer of biocompatible shell on the quantum dots, which could be polymers, non-toxic organic materials etc.

With the progress of nanomaterials and nanotechnology, researchers are now able to synthesize less toxic fluorescent nanomaterials for biomedical imaging. These fluorescent materials includes carbon-based quantum dots such as carbon quantum dots, graphene quantum dots etc. and noble metal nanoclusters such as Au, Pt, Cu nanoclusters etc. [84–87]. For instance, Wu et al. and Zhao et al. first use cocoon silk as the precursor to prepare fluorescent carbon nanoparticles through a

**Fig. 1.9** Fluorescent QDs with different emission wavelength [83]

hydrothermal method. The resulting carbon nanoparticles have a high quantum yield of 38% and high photostability. Cell experiments have proven that rapid fluorescent cell labeling could be achieved with these carbon nanoparticles, as shown in Fig. 1.10 [88].

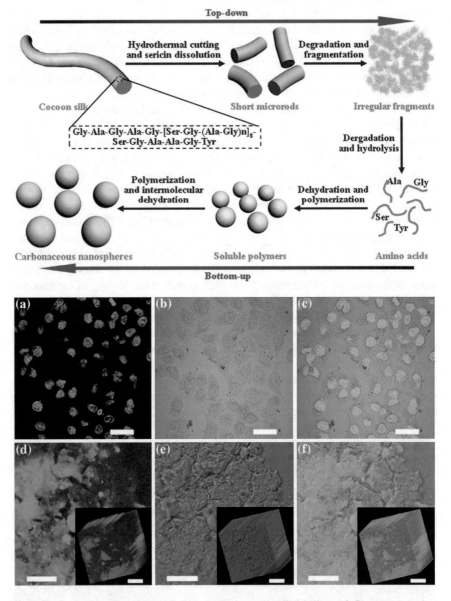

**Fig. 1.10** Schematic illustration of the synthesis and cell labeling of fluorescent carbon nanoparticle with the cocoon silk as the precursor [88]. (Copyright 2013 Wiley-VCH)

From the principle of emission spectrometry, the fluorescent nanomaterials mentioned above are all classified as downconversion fluorescence, which relies on using high energy photon to excite the emission of low energy photon. Their excitation and emission wavelength are generally in the regions of ultraviolet or visible light. When applied to fluorescent bioimaging, ultraviolet light causes photodamage to human body, and the penetration depth of visible light is very limited. Thus, near-infrared upconversion fluorescent nanoparticles gradually catch people's attention [88–95]. Upconversion fluorescent nanoparticles are mainly composed of the lanthanide metals. Unlike downconversion fluorescent nanoparticles, upconversion fluorescent nanoparticles use low energy near-infrared light as excitation source to emit high energy photons. Furthermore, by doping the upconversion fluorescent nanoparticles with other lanthanide metal ions, the peak of emission can be brought to the near-infrared region, and hence near Infrared-near-infrared upconversion fluorescence is achieved. The biggest advantage of near-infrared fluorescence in fluorescent imaging is that it can achieve centimeter-level penetration depth.

### 1.3.2.2 Application of Nanoprobes in Magnetic Resonance Imaging

Magnetic resonance imaging (MRI) is one of the most widely used molecular imaging techniques in clinical diagnosis nowadays. Its imaging principle is based on the change in energy state of hydrogen nuclei in the human body under radio frequency field, where it will return from the excited state to the initial energy state as the radio frequency is removed, and this process is denoted as relaxation. During relaxation of the hydrogen protons, magnetic resonance signal is accepted through signal reception, processing and computer reconstruction, and a MR image reflecting the 3-D structure of human body can thus be obtained. The strength of MR signal depends on the density of the hydrogen nuclei, which varies in different types of human tissues. For instance, there is a difference in moisture between normal tissue and pathogenic tissue, which could be reflected by MRI [96–98]. In early MRI application, people did not realize the importance of contrast agent. However, after a period of development, medical researchers discovered between benign and malignant tumors, certain normal and diseased tissue have weaker MR signal difference because of the overlap of relaxation time of T1 or T2. This makes it difficult to diagnose since normal and diseased tissue cannot be distinguished effectively. It is also hard to detect small volume of diseased tissue, thus missing the optimal time for treatment. Therefore, MRI contrast agents have emerged. MRI contrast agents by themselves do not produce any MR signal, yet they could change the relaxation rate of water protons in the surrounding tissues, thereby enhancing the contrast between normal and diseased tissues. According to the imaging mechanisms, MRI contrast agents can be divided into two main categories: T1 and T2 MRI contrast agents.

The use of nanomaterials in MRI can be traced back to 1978, when it was first discovered that nanomaterials can change water proton relaxation time. Yet, the use of nanomaterials in MRI contrast agents starts in 1986. With the development of nanoscience, the variety of nanoparticle-based MRI contrast agents rapidly increases, the techniques for controlled synthesis have been greatly improved, and the relaxation performance has been enhanced. According to the imaging principle, the currently available nanoparticle-based MRI contrast agents can be classified into two categories:

1. Nanoparticle-based T1 MRI contrast agents

T1 MRI contrast reagents are mainly based on paramagnetic substances, such as paramagnetic ions like $Gd^{3+}$, $Mn^{2+}$, $Fe^{3+}$ etc. These substances can shorten the T1 relaxation time of hydrogen protons, leading to enhanced magnetic resonance signals. Currently, the clinically used T1 contrast agents are mainly Gd-based organic complexes, such as Magnevist (a chelate formed from the combination of $Gd^{3+}$ ions and diethylenetriamine pentaacetic acid, Gd-DTPA). Nevertheless, after 30 years of clinical application, the shortages of this type of organic gadolinium complexes gradually reveal, mainly as (1) extremely short in vivo circulation time; it will be excreted from the blood circulation system soon after intravenous injection, and its concentration in blood will be lowered by 70% in 5 min. Having short half-life in blood is unfavorable for some applications that require long-term circulation; (2) Side effects. Free $Gd^{3+}$ ion is very toxic, but its toxicity is significantly reduced once it forms a chelate with ligands. Some studies stated that gadolinium ion leakage is observed in clinically used gadolinium complex, causing strong side effects. Recently, the US FDA issued a warning about the use of gadolinium-based contrast agents may lead to nephrogenic systemic fibrosis (NSF). (3) Lack of specificity, targeting imaging is thus impossible. Although targeting ligands can be conjugated to the complex via conventional interaction, the resulting contrast agent still faces the problem of poor stability, short half-life etc.

In order to overcome the shortages of these small molecular Gd-based contrast agents, increasing efforts have been devoted to the development of nano-sized MRI contrast agents containing paramagnetic ions [99–105]. A simple approach is to graft Gd-DTPA onto the surface of nanomaterials by covalent or non-covalent interaction, thereby extending the in vivo circulation time. For instance, Mulder et al. prepared organic ligands containing Gd-DTPA, and these ligands were attached to the surface of various nanoparticles through hydrophobic interactions, as shown in Fig. 1.11 [106]. Another example is, in 2007, Lin et al. covalently linked Gd-DTPA to amino group-containing silane molecules, and these complexes were further attached to the surface of biocompatible silica by hydrolysis (Fig. 1.12), and the nanoprobe can be used for imaging of tumor cells [107].

In order to prevent the leakage of gadolinium ion, researches tend to synthesize stable nanomaterials containing gadolinium ions. For example, Riviere et al. and Roux et al. developed a strategy for controllable synthesis of silica-coated gadolinium oxide nanoparticles. Compared to the Gd-DTPA contrast agent in clinical use, gadolinium oxide nanoparticles have higher relaxivity [108]. Furthermore, the nature of gadolinium nanoparticle itself and the silica shell can

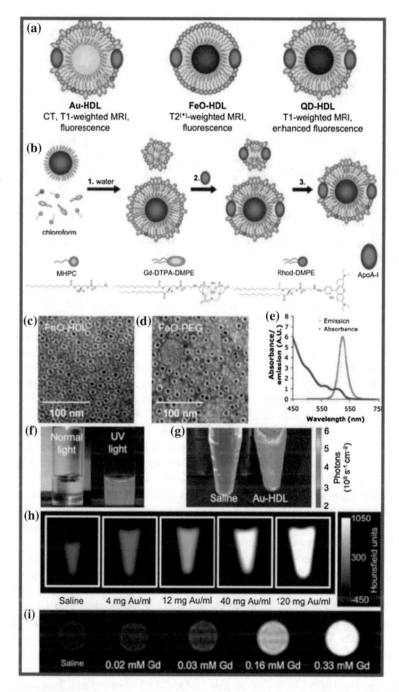

**Fig. 1.11** Covalent conjugation of Gd-chelate on the surface of CNT [106]. (Copyright 2008 American Chemical Society)

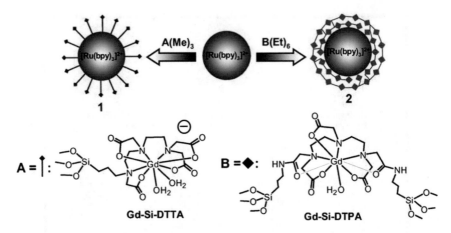

**Fig. 1.12** Covalent conjugation of Gd-chelate on the surface of silica nanoparticles [107]. (Copyright 2007 Wiley-VCH)

effectively prevent leakage of gadolinium ions. In addition, many research groups use Gd as the upconversion fluorescent host material to synthesis Gd-containing upconversion fluorescent nanoparticles [109, 110]. These nanoparticles can be used for simultaneous MRI and fluorescence imaging. However, the relaxation of Gd-containing upconversion fluorescent nanoparticles is much weaker than that of clinical Gd-DTPA. In order to enhance the relaxation performance of these materials, researchers propose that this could be achieved by increasing Gd concentration on the nanoparticle surface and local surface water molecule concentration. For example, Chen et al. fabricated $NaYF_4:Er/Yb/NaGdF_4$ nanoparticles and coated them with a layer of mesoporous silica (Fig. 1.13). Using these nanoparticles as T1 contrast agent, its longitudinal relaxivity was significantly increased [111].

Another effective way to prevent gadolinium ions leakage is using other paramagnetic ions as T1 contrast agent. Recently, the use of paramagnetic $Mn^{2+}$ ions as the T1 contrast agent has been extensively studied [112–114]. In 2007, Hyeon et al. first reported the use of manganese oxide as the T1 contrast agent [113]. Through adjusting the reaction conditions, manganese oxide nanoparticles with different sizes were synthesized. After modification with amphiphilic molecules, manganese oxide nanoparticles were successfully applied to MR imaging of the brain in mice. As shown in Fig. 1.14, the brain MR signal significantly increases upon injection of manganese oxide nanoparticles. Nevertheless, the biggest disadvantage of using $Mn^{2+}$ ions as T1 contrast agent is low relaxivity performance, which is usually less than $1 \, mM^{-1}S^{-1}$.

Apart from $Mn^{2+}$ ions, recent literature has reported the use of $Cu^{2+}$ ions as T1 contrast agent. For example: Pan et al. from University of Washington prepared self-assemble $Cu^{2+}$ ions-containing nanoparticles and applied it to in vivo T1-weighted imaging [115]. Experiments have demonstrated that the T1 relaxivity of this contrast agent is similar to that of Gd-DTPA, but it has a longer in vivo circulation time. This contrast agent can be used to image thrombosis in rats after intravenous injection.

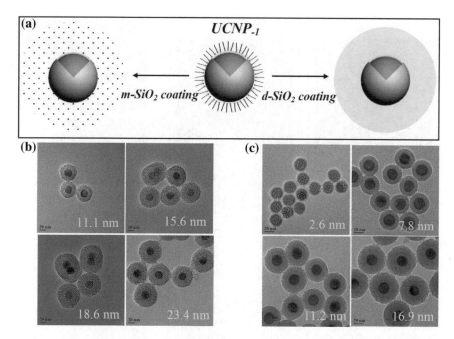

**Fig. 1.13** Silica-coated upconversion fluorescence nanoparticles for MRI [111]. (Copyright 2013 Wiley-VCH)

## 2. Nanoparticle-based T2 MRI contrast agents

The use of nanoparticle-based MRI contrast agents for T2-weighted imaging is contributed to the presence of magnetic nanomaterials. The most typical material is the superparamagnetic iron oxide [116, 117]. Compared to Gd- and Mn-based contrast agents, iron oxide has high biocompatibility, and no significant toxic side effects were observed upon intravenous injection into the human body. Currently, D-dextran-coated iron oxide nanoparticles are in clinical use. Its principle is different from that of the T1 contrast agents. When an external magnetic field is introduced, the magnetic nanoparticles would lead to a non-uniform magnetic field, and proton transverse magnetization is observed when water molecules pass through this non-uniform magnetic field. This decreased T2 relaxation time of protons can also be interpreted as increased T2 relaxivity performance. Nevertheless, T2 images show a dark contrast compared to T1 images, which easily causes confusion with human physiological phenomena such as calcification, and hence leads to misdiagnosis. Theoretically, $Fe^{3+}$ ions can be used as a T1 contrast agent, yet, this functionality is hindered by strong magnetic property of iron oxide. In 2011, Hyeon et al. fabricated ultrasmall iron oxide nanoparticles. Because of reduced dimension, the magnetic moment of the nanoparticles drastically reduced, which enables iron oxide to be used as a T1 contrast agent with a relaxation rate comparable to that of Gd-DTPA [118].

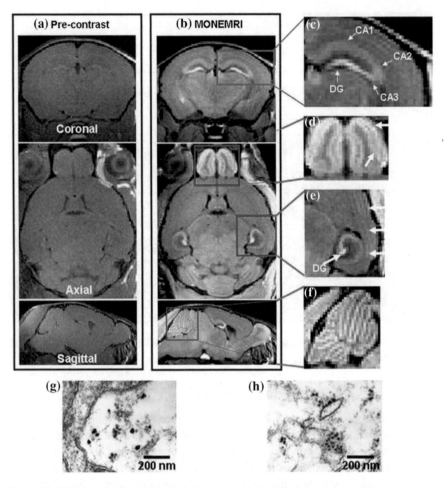

**Fig. 1.14**  MRI images of mouse brain before vs after injection of manganese oxide nanoparticles [113]. (Copyright 2007 Wiley-VCH)

### 1.3.2.3  Application of Nanoprobes in X-Ray CT Imaging

X-ray CT imaging is also known as X-ray computed tomography imaging, and its principle is illustrated in Fig. 1.15, when X-ray penetrates through the body, the extent of X-ray adsorbed varies from tissue to tissue, resulting in differences in CT signal [119]. The large difference in attenuation coefficient allows one to distinguish skeletal tissue and cartilage tissue easily. Nevertheless, the X-ray attenuation coefficient between different soft tissues is closed, which makes it hard to

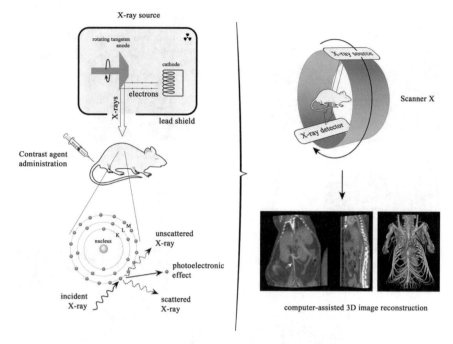

**Fig. 1.15** Mechanism of X-ray CT imaging [119]. (Copyright 2012 Wiley-VCH)

distinguish. Thus, injection of contrast agent into the patient is needed for clinical CT scan for soft tissue. Currently, clinical X-ray CT contrast agent mainly composes of iodine-substituted aromatic organic small molecules. Yet, these small molecules have a lot of shortcomings: such as weak contrast ability; short in vivo circulation time, lack of targeting, risk of kidney failure, among others [120].

To overcome the drawbacks of clinical CT contrast agents, researchers have investigated various nanoparticle-based CT contrast agents, which can be classified into two main categories: iodine-containing organic polymers and inorganic nanoparticles. The typical method to fabricate iodine-containing organic polymeric nanoparticles such as polylactic acid is to package iodine-containing small molecules into organic polymeric nanoparticles through covalent or non-covalent interaction [121]. Various inorganic nanoparticles have been studied as CT contrast agents.

1. Gold nanoparticles. Owing to the large attenuation coefficient of Au, and various gold nanostructures including gold nanosphere, gold nanorod, and gold nanostar etc. (Fig. 1.16) have been investigated [122–125]. Another reason why gold nanoparticles are used as CT contrast agents is because of their high chemical stability and low toxicity. For example, Jon et al. fabricated PEGylated gold nanoparticles and applied them as X-ray blood-pool CT contrast agent [122]; Eck et al. used antibody-conjugated nanoparticles for CT imaging of lymph nodes [123]; Kim and Ahn et al. used heparin-DOPA as the stabilized agent to fabricate gold nanoparticles, and these nanoparticles can be used for

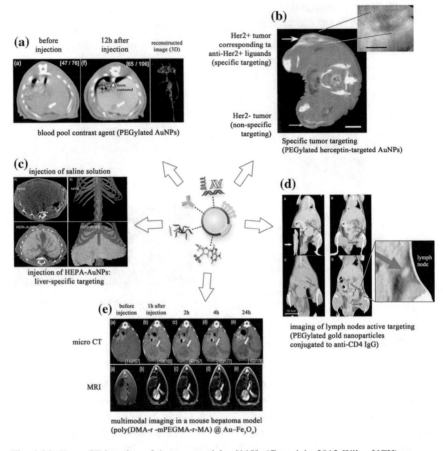

**Fig. 1.16** X-ray CT imaging of Au nanoparticles [119]. (Copyright 2012 Wiley-VCH)

targeted imaging of the liver [124]. In addition to gold nanoparticles, noble metal X-ray CT contrast agents also include platinum and silver nanoparticles [125]. Despite the high quality of the in vivo imaging, CT imaging requires a large volume of the contrast agent, and these metals are expensive, which restrict their application in clinical imaging.

2. Bismuth sulfide nanoparticles. In 2006, Weissleder et al. initially used bismuth sulfide nanoparticles as the CT contrast agent, and these bismuth sulfide nanoparticles displayed high contrast [126]. Nevertheless, the controlled synthesis and surface modification of bismuth sulfide remains challenging and thus limits the application of bismuth sulfide in X-ray CT imaging.

3. Tantalum oxide nanoparticles. In order to reduce the cost of contrast agents and to achieve controlled synthesis, in 2011, Choi and Hyeon et al. fabricated tantalum oxide nanoparticles (Fig. 1.17), and these nanoparticles were successfully used for in vivo CT imaging and lymph node imaging in rat [127].

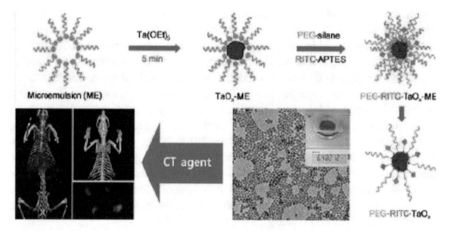

**Fig. 1.17** X-ray CT imaging of TaOx nanoparticles [127]. (Copyright 2011 American Chemical Society)

### 1.3.2.4 Application of Nanoprobes in Ultrasound Imaging

Ultrasound imaging technology emerged in the 1950s, and became widely used in clinical examination in the 1970s. Ultrasound imaging is based on the Doppler effect. Different tissues in the human body have different acoustic impedance, when the difference in acoustic impedance of two sound waves is greater than 1/1000, acoustic interface is formed. The incident ultrasonic wave would be reflected or scattered when passes through the acoustic interface, detected by the detector, and then processed to form an image. Human organs can be classified into four categories according to the echogenicity echo characteristics of different tissues:

1. Non-reflective: This category includes blood, amniotic fluid, and urea etc. Since these liquids are uniform, they do not have significant difference in acoustic impedance, and the reflection coefficient is 0;
2. Low-reflective: This category mainly includes soft tissues, and the acoustic impedance is low;
3. High-reflective: Parenchyma tissue has larger difference in acoustic impedance, thus its reflection coefficient is also larger;
4. Total-reflective: The acoustic interface between soft tissue and gas-containing tissue has the greatest acoustic impedance, the reflection coefficient can thus reach up to 99.9%

Similar to MRI and CT imaging, ultrasound imaging also needs contrast agents. As indicated by the reflection coefficient, the echo formed between gas and tissue is the strongest, and therefore most of the ultrasound contrast agents contain gas bubbles. Nevertheless, gases diffuse fast and are short-living. Nanoparticle-based contrast agents have thus been introduced in ultrasound imaging. To date, the major categories of nanoparticle-based ultrasound contrast agents include: (1) phospholipids microbubbles contrast agent; (2) inorganic encapsulated nanoparticles (e.g., hollow silica nanocapsules); (3) perfluorocarbon nanoparticles; this type of contrast agent has an advantage compared to other contrast agents. Since gases diffuse fast, researchers turn their focus onto liquid media. However, the reflective coefficient of liquid is small, and the echo is thus weaker. After continuous exploration, researchers focus their attention on liquid that has low boiling point and are biocompatible. This liquid can be packaged into shells, and would transform into gas and generate a strong echo under ultrasound. The most commonly used liquid is perfluorocarbon molecules, such as perfluorohexane. For example, in 2012, Wang et al. fabricated hollow mesoporous silica nanocapsules, and then encapsulated perfluorohexane into the nanocapsules simply by sonication [128–130]. Under ultrasound, perfluorohexane would be gasified, and can be detected by ultrasonic testing, as demonstrated in Fig. 1.18. Compared to clinical ultrasound contrast agents, this type of contrast agent has longer shelf life, and thereby strong signal stability [131].

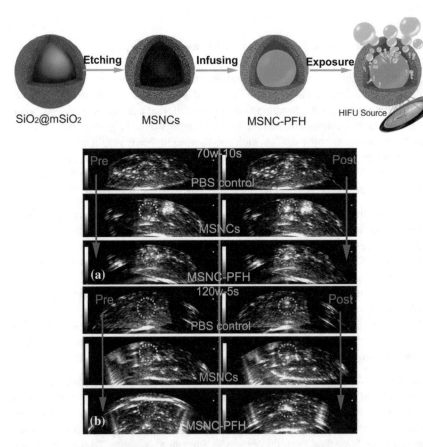

**Fig. 1.18** Synthesis of ultrasound imaging contrast agents and their ultrasound images [128]. (Copyright 2003 Wiley-VCH)

### 1.3.2.5   Application of Nanoprobes in Photoacoustic Imaging

Photoacoustic imaging is a novel noninvasive medical diagnostic technique developed in recent years, it is a combination of optical and ultrasound technology. In photoacoustic imaging, when light pulse shines onto a biological tissue, the tissue would absorb the light and convert light energy into thermal energy, causing thermoelastic expansion in the surrounding tissue and generating ultrasonic waves. These waves can be detected by an ultrasound transducer, and by measuring the photoacoustic signal, the light absorption distribution image of the tissue can be reconstructed, achieving detection purposes. Although photoacoustic imaging has not been applied to the clinical, this technology combines the features of high contrast of pure optical imaging and high penetration of pure ultrasound imaging, generating tissue images with high contrast and high resolution, thus is a very promising medical diagnostic technology. Nanoparticles photoacoustic contrast agents have been booming in recent years, a variety of inorganic nanoparticles and organic polymeric nanoparticles demonstrate high quality photoacoustic imaging. These materials mainly includes carbon nanomaterials such as carbon nanotubes, noble metal nanomaterials such as gold and silver nanoparticles, inorganic semiconductors such as copper sulfate, organic polymeric nanoparticles etc. [132–135]. For example, Gambhir et al. modified the surface of single-walled carbon nanotubes with PEG and targeting protein to investigate the performance of carbon nanotubes in vivo photoacoustic molecular imaging. As shown in Fig. 1.19, through adjusting the molecular mass of PEG and changing the targeting ligand, single-walled carbon nanotubes can effectively concentrate at the tumor site, achieving targeted diagnosis of tumors [136].

### 1.3.2.6   Application of Nanoprobes in Nuclear Imaging

The two major types of nuclear imaging are positron emission tomography (PET) and single photon emission computed tomography (SPECT). Both of these techniques are based on tracking the distribution of radionuclides in the human body to obtain information on human anatomy, and they have high sensitivity. Nevertheless, radioactive elements are costly and harmful to the human body. Generally speaking, the development of nanomaterials in PET and SPECT imaging is not as rapid as in other imaging techniques. Currently, the most widely used radioactive element in PET imaging is $^{64}$Cu. It is grafted onto the nanoparticle surface by chelation with other ligands. For instance, Cai et al. modified the surface of graphene with NOTA and targeting ligand, a tumor targeting PET imaging probe is constructed taking the advantage of strong coordination ability between $^{64}$Cu and NOTA [137]. Xia et al. modified the surface of gold nanocages with $^{64}$Cu-NOTA ligand and use PET imaging technique to examine the pharmacokinetics of gold nanocages in living mice [138]. In order to prevent $^{64}$Cu falling off from the ligand, Prof. Louie from the University of California doped $^{64}$Cu into iron oxide, achieving PET/MRI dual-mode imaging [139].

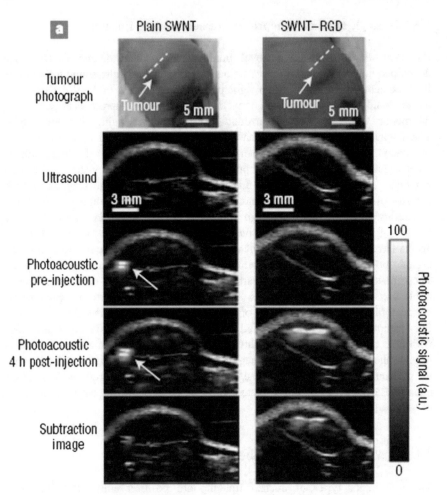

**Fig. 1.19** Photoacoustic imaging of SCNT [136]. (Copyright 2008 Nature)

### 1.3.2.7   Trends in Development of Nanoprobes in Molecular Imaging

Early diagnosis not only can significantly enhance the cure rate of certain diseases, but also can reduce patient's pain and treatment costs. From this perspective, currently available single-mode contrast agents can no longer meet people's need. Each imaging modality has its own advantages and limitations. Table 1.1 summarizes the advantages and disadvantages of different imaging modalities. As illustrated in the table, certain imaging modalities complement each other. Therefore, more accurate diagnostic information can be obtained by combining different imaging modalities, which aid overcoming the limitations of single-mode imaging. Multi-mode imaging has aroused people's attention, and has been widely developed in the past few years. Up to date, numerous reports on combination of

**Table 1.1**  Advantages and disadvantages of different imaging tools [140]

| Imaging technique | Disadvantages | Advantages | Possibility of human imaging |
|---|---|---|---|
| PET -SPECT | Low spatial resolution, radiation risks, high cost (for PET, cyclotron or generator needed) | High sensitivity, quantitative, no penetration limit | Yes |
| CT | Not quantitative, radiation risks, limited soft tissue resolution, limited molecular applications | Anatomical imaging, bone and tumor imaging | Yes |
| MRI | Low sensitivity, high cost, time consuming scan and processing | Morphological and functional imaging, no penetration limit, high spatial resolution | Yes |
| Optical imaging | Photobleaching, limited penetration, low spatial resolution; autofluorescence disturbing | Low cost, easy manipulation, high sensitivity, detection of fluorochrome in live and dead cells | Yes but limited |
| US | Limited resolution and sensitivity, low data reproducibility | Safety, low cost, wide availability, real time | Yes |

different imaging modalities have been published, such as MRI/fluorescence, MRI/CT, SPET/CT, CT/optical imaging etc. [141–145]. In 2008, Bruns et al. published an article in *Nature nanotechnology* on the application of combination of MRI and fluorescence imaging in in vivo imaging (Fig. 1.20). Bruns et al. encapsulated iron oxide and semiconductor quantum dots simultaneously into lipoprotein nanocrystals. Using this nanoprobe not only improve the quality of the images, but also allows one to observe the endocytosis of lipoprotein nanocrystals and in vivo metabolism condition quantitatively and dynamically [145].

## 1.3.3  Application of Nanoprobes in Cancer Treatment

### 1.3.3.1  Types of Clinical Cancer Treatment

There are many cancer therapies available nowadays, such as biological therapy, but to date, clinical cancer treatments still rely heavily on surgery, radiotherapy, and chemotherapy. The following sections would discuss these three treatments in depth:

1. Surgery

Surgery, as suggested by its name, is a physical intervention for complete or partial removal of tumor tissue, and is the oldest clinical cancer treatment. The advantages of surgical cancer treatment are direct and quick, and surgery is thus the first choice

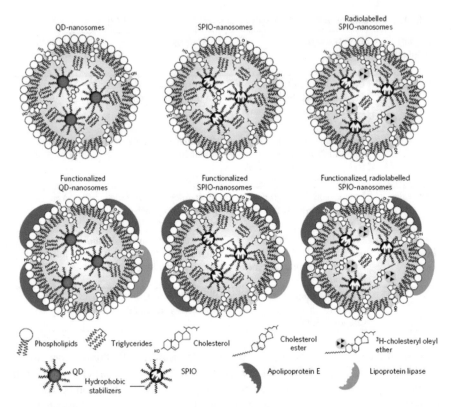

**Fig. 1.20** Schematic illustration of the MRI-fluorescence dual model imaging nanoprobes [145]. (Copyright 2009 Nature)

for early-stage and mid-stage cancer treatment. Even for lung cancer, which has a high mortality rate, if tumor is removed in early stage by surgery, the cancer cure rate can reach 50–70%. Many people think that all the tumors should be removed. However, surgical resection is not suitable for all cancer patients. There are lots of problems associated with surgical treatment, and the surgical risk is extremely high for some sensitive cancer. The currently available surgical treatments for cancer are mainly classified into radical surgery and palliative surgery. Surgical resection is more traumatic to patients, and some are difficult to recover, which weaken the immunity of the patients and hence more prone to a series of complications. Furthermore, for small lesions or metastatic disease, and especially those in the later stage of cancer where tumor widely spread or invades major organs and major blood vessels, surgical resection is no longer effective.

2. Radiation therapy

Radiation therapy or radiotherapy is a therapy which relies on the use of high-energy electromagnetic radiation including α-, β-, and γ-ray generated by radioisotopes, therapeutic X-ray generated from accelerator, and electronic beam, proton beam and other particle beam generated from various types of accelerator, to kill malignant tumor cells. Soon after the discovery of X-ray by Roentgen and the discovery of radium by Madame Curie, radiotherapy was put into use. This method has played an important role in cancer therapy for more than a century, and is still one of the most important methods for cancer therapy. From statistical result, about 70% of cancer patients require radiotherapy. Mechanism of radiotherapy includes direct damage and indirect damage. Direct damage refers to the situation where high-energy radiation acts directly on the organic molecules, producing large amount of free radicals, which causes crossing over of breakage of DNA molecules. Indirect damage refers to the ionization of water molecules in human body caused by high-energy radiation, producing large amount of free radicals, and these free radicals further react with organic molecules, leading to irreversible damage to the targeted tissues. Although radiotherapy can kill tumor cells, it can also damage normal cells, inducing the incidence of the second cancer.

3. Chemotherapy

Chemotherapy is a type of cancer therapy that involves the use of synthetic drugs to kill or suppress the growth of tumor cells. Chemotherapy is a systemic treatment. Typically, chemotherapy drugs can be given by oral, intramuscular or subcutaneous injection, intrathecal injection, and intravenous injection routes. Thus, this treatment has therapeutic effect to primary tumors, metastases and subclinical metastases. For patients with late-stage cancer, the cancer has spread over a large area to a point where surgery and radiation therapy can no longer control metastasis, and chemotherapy becomes the major treatment method. Currently, there are many types of chemotherapy drugs. According to their chemical structure and working principle, chemotherapy drugs are classified into groups such as alkylating agents, herbal, hormonal and antibiotics etc. Although chemotherapy is effective for primary lesions and metastatic lesions, all the chemotherapy drugs are toxic to normal cells and lack of targeting ability, thus it is impossible to prevent normal cells from being damaged. Furthermore, chemotherapy drugs have strong side effects including hair loss, nausea, vomiting, weakening in immunity etc. Due to environmental pollution, poor diet, and other factors, the incidence of cancer is increasing year by year, and so the development of new treatment methods and effective therapeutic agents is highly desirable. The use of nanomaterials in cancer therapy is one of the hot research topics in nanomedicine. With the development of nanotechnology and novel nanomaterials, numerous efforts have been made to improve the therapeutic efficacy of cancer. The following lists a few therapeutic methods that involve the extensive use of nanomaterials:

### 1.3.3.2 Application of Nanomaterials in Controlled Drug Release

As mentioned above, anti-cancer drugs are non-targeting, and will accumulate in both diseased and normal tissues, causing severe side effects. Nanomedicine has demonstrated tremendous opportunities to overcome these issues by significantly improving the pharmacokinetics of the drug. Up to date, a variety of nanomaterials such as mesoporous silica nanoparticles, polymeric capsules, carbon-based nano-materials etc. have been investigated as drug carriers [146–150]. Furthermore, by conjugating targeting ligands to the surface of the nanoparticles and using the external stimulus such as temperature, pH, light etc., targeted delivery and con-trolled release of anti-cancer drugs can be achieved. For example, Zhang et al. utilized mesoporous silica as drug carriers, by decorating the surface of silica with β-cyclodextrin via disulfide bonds. The cavity of β-cyclodextrin is hydrophobic, which enables a host-guest interaction with the targeting protein RGD and the PLGVR peptide (Fig. 1.21). The resulting nanomaterial showed high tumor accu-mulation. Moreover, the glutathione in the tumor cells would break the disulfide bonds between silica and β- cyclodextrin, and hence drug is released [151]. In another study, Zhang et al. coated a mesoporous silica shell onto Pd/Ag alloy, and then modified the silica shell with 3, 4-hydroxybenzaldehyde to coordinate with $Fe^{3+}$ ions. Meanwhile, the anti-cancer drug doxorubicin (DOX) could be loaded within the silica shell. With this method, the drug loading efficiency of DOX is 49%, and DOX could be released from the silica shell under acidic pH. In the meantime, Pd @ Ag alloy absorbs light and convert it into thermal energy, to further promote the release of DOX under near-infrared light irradiation [152]. Lin et al. encapsulated upconversion nanoparticles into a mesoporous silica shell, and further modified the surface of silica with a layer of heat-responsive polymer.

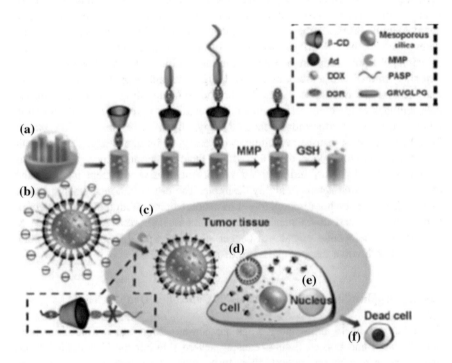

**Fig. 1.21** Schematic illustration of controllable drug delivery based on mesoporous silica nanoparticles [151]. (Copyright 2013 American Chemical Society)

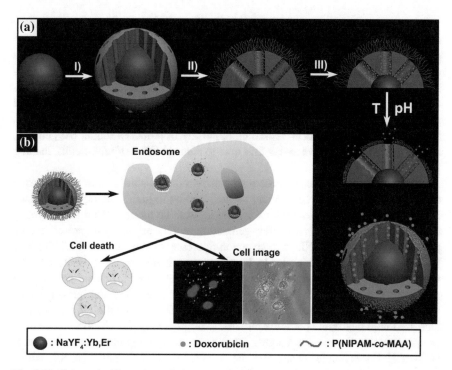

**Fig. 1.22** Schematic illustration of thermal and pH-responsive drug delivery vehicles [153]. (Copyright 2013 Wiley-VCH)

Doxorubicin was then loaded into the silica pores (Fig. 1.22). This system is highly stable under low temperature and physiological pH, and thus low drug release. Nevertheless, when the temperature is raised or the pH is lowered, the drug is rapidly released from the silica pores, indicating the controlled drug release [153].

### 1.3.3.3 Application of Nanomaterials in Photodynamic Therapy

4000 years ago in the period of ancient Egypt, people discovered when patients with vitiligo took psoralen extracted from plant orally, psoralen would accumulate under the skin, and the white patches would disappear after exposure to sunlight. Not until in 1976 the clinical application of a porphyrin derivative successfully treated a bladder cancer patient, leading to the development of photodynamic therapy for cancer. Photodynamic therapy (PDT) is a non-radioactive, minimally invasive treatment method. The basic elements of photodynamic therapy include oxygen, photosensitizer, and laser. First, upon exposure to light with specific wavelength, the photosensitizer will be activated and undergo photochemical reactions to generate highly active $^1O_2$ to kill tumor cells. The effectiveness of photodynamic therapy in some cancers is comparable to surgery, chemotherapy, and radiotherapy, and may even fully treat some early-stage cancers.

The most critical step of photodynamic therapy is to make photosensitizer accumulate effectively at the tumor site. To achieve this goal, photosensitizers are generally encapsulated or covalently conjugated to the nanoparticles, given the prolonged in vivo circulation time and ease of modification of nanoparticles [154–160]. For example, Dai et al. developed a novel photosensitive agent by covalently conjugating porphyrin molecules to the lipid-containing amphiphilic polymer, and produced porphyrin-containing polymeric nanoparticles by a self-assembly process, as shown in Fig. 1.23. Under visible light, porphyrin produces singlet oxygen $^1O_2$, and $^1O_2$ is highly oxidative and is thus able to damage DNA of tumor cells, leading to cell death [160].

Visible light is known to have poor penetration capability, due to the strong absorption of visible light by human tissues, leading to limited therapeutic efficacy coupled with high phototoxicity to normal tissues. To address these issues, many research groups have reported a new generation of PDT agents by incorporating the photosensitizers into the NIR upconversion fluorescent nanoparticles (UCNPs). The UCNPs nanoparticles emit visible light upon excitation by NIR laser, and the visible light is then absorbed by the photosensitizers for PDT. This process can efficiently avoid the direct exposure of tissues to visible lights and human tissues have low absorption in the NIR region, and thus the therapeutic efficacy can be increased with negligible phototoxicity. More impressively, the UCNPs can be used as upconversion fluorescent imaging for potential imaging-guided cancer treatment [161–165]. For example, Zhang et al. have coated the NaYF$_4$:Yb/Er UCNPs with a mesoporous silica shell, and encapsulated ZnPc within the mesoporous silica shell.

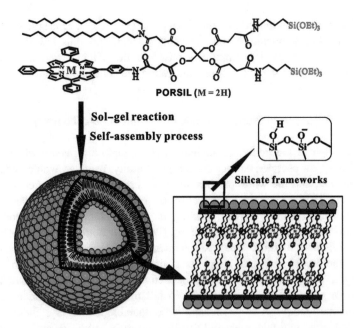

**Fig. 1.23** Photosensitizer based on porphyrin-containing polymeric nanoparticles [160]. (Copyright 2011 Wiley-VCH)

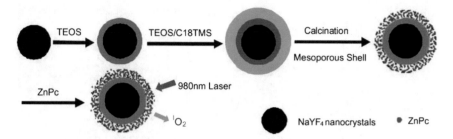

**Fig. 1.24** PDT agent based on ZnPc-loaded mesoporous silica-coated UCNPs [165]. (Copyright 2009 Wiley-VCH)

Upon irradiation with a 980 nm laser, the resulting NaYF$_4$:Yb/Er UCNPs emitted the fluorescence at 650 nm, which was further absorbed by ZnPc to generate $^1O_2$ for killing cancer cells [165] (Fig. 24).

### 1.3.3.4   Applications of Nanomaterials in the Photothermal Therapy

Photothermal therapy (PTT) employs photoabsorbing agents to locally convert the light into heat to induce cell apoptosis in tumor. PTT has high selectivity, as it will induce cell apoptosis only with PTT agent accumulation followed by laser irradiation. PTT is also considered to be safe because the NIR light generally used for PTT has low absorbance by human tissues. Therefore, PTT has recently become a fast-growing and effective tool for safe and effective treatment of cancers. In principle, materials that have high absorbance in the NIR region can be used for PTT agents. Conventional PTT agents are NIR dyes, but they have poor photostability and water-solubility, and showed modest therapeutic efficacy. To overcome these obstacles, various nanoparticle-based PTT agents have been developed, which are roughly divided into the following categories:

1. Gold nanomaterials

It has long been recognized that gold nanomaterials have unique optical properties, and are easy to prepare and modified with functionalities. Various gold nanostructures including gold nanoshperes, gold nanorods, gold nanocages, and gold nanoshells have been studied as PTT agents [166–170]. Tae's group has used chitosan-coated gold nanorods for PTT of cancer [166]. Farokhzad's group developed gold nanorods modified with DNA and targeting ligand through a self-assemble method for targeted PTT therapy of cancer [167]. oo et al. have developed gold half-shell multifunctional nanoparticles by modifying targeted ligands on the surface and simultaneously loading the chemotherapeutic drug for synergistic cancer therapy, as shown in Fig. 1.25 [168].

2. Carbon nanomaterials

Carbon nanomaterials, in particular carbon nanotubes and graphene, have large specific surface area and high conductivity, and have received great attention in the

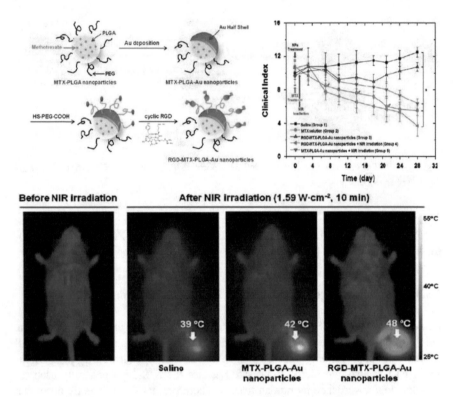

**Fig. 1.25** The synthesis and application of gold half-shell multifunctional nanoparticles in the cancer treatment [168]. (Copyright 2013 American Chemical Society)

field of energy. In addition to drug delivery, carbon nanotubes and graphene have strong absorbance in the NIR region, and thus have been investigated as PTT agents [170–175]. Given the poor water-solubility of carbon nanotubes, amphiphilic PEG molecules have been modified on the surface of carbon nanotubes to increase the water-solubility. Under laser irradiation with a 808 nm laser, PEG-modified carbon nanotubes efficiently killed the cells through NIR-mediated hyperthermia. In addition, they showed long circulation time in the blood, and about 15% of the nanotubes remained in the blood at 48 h postinjection. Efficient tumor inhibition was observed after laser irradiation [176]. Compared to carbon nanotubes, evidence has further confirmed that graphene has higher photothermal conversion efficiency (Fig 1.26).

3. Semiconductor nanomaterials

Semiconductor nanomaterial-based PTT agents are mainly focused on the copper-based semiconductor nanomaterials [177–181]. In 2011, Hu's group has developed novel flower-like CuS nanoplates. These nanoplates also have strong absorbance at 980 nm, and could kill the cancer cells upon laser irradiation [180]. Compared to the aforementioned PTT nanomaterials, a 980 nm laser was used for CuS nanoplates, which has deeper penetrating capability compared to 808 nm light. Nevertheless,

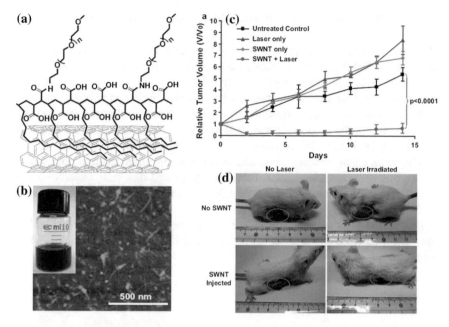

**Fig. 1.26** The synthesis and application of CNT for the PTT [176]. (Copyright 2011 Elsevier)

these materials have also shown some limitations such as lower photothermal conversion efficiency, large particle size, and poor dispersibility. In the same year, Brian A. Korgel et al. reported hydrophobic $Cu_{2-x}Se$ nanocrystals with the uniform size through a colloidal hot injection method [181]. After coating with an amphiphilic polymer, The nanocrystals readily disperse in water and exhibit strong near-infrared (NIR) optical absorption with a high molar extinction coefficient of $7.7 \times 10^7$ cm$^{-1}$ M$^{-1}$ at 980 nm. When excited with 800 nm light, the $Cu_{2-x}Se$ nanocrystals produce significant photothermal heating with a photothermal transduction efficiency of 22%, comparable to gold nanorods and nanoshells [182–185].

4. Polymer-based nanomaterials

Polymer-based nanomaterials have been considered "soft" materials compared to the inorganic nanomaterials. Generally, polymer nanomaterial-based PTT agents exhibit lower toxicity and are biodegradable, as compared with "hard" inorganic nanomaterials. Until date, various polymer nanoparticle-based PTT agents have been studied and demonstrated great potential in cancer therapy [186–190]. Haam's group have reported a polyaniline nanoparticle-based PTT agent, and the temperature of the polyaniline nanoparticle solution increased by $\sim 40$°C upon irradiation with a 808 nm laser for 3 min [190]. Poly pyrrole nanoparticles have also been developed by Liu et al. as a PTT agent. To further decrease the toxicity, Zheng et al. have synthesized porphysome nanovesicles using porphyrin as the unit. The porphysome nanovesicles demonstrated excellent biocompatibility and enzyme-mediated biodegradation [191, 192].

### 1.3.3.5  Application of Nanomaterials in Magnetic Hyperthermia

Magnetic hyperthermia is an experimental treatment for cancer. It is theoretically based on the fact that magnetic nanoparticles can transform electromagnetic energy from an external high-frequency field to heat. The major superiority of this strategy is selectivity and targeting. The magnet can increase the nanoparticle accumulation in the target site and thus increase the therapeutic efficacy. Moreover, the magnetic nanoparticles can be used as T2-weighted MRI imaging for simultaneous diagnosis and therapy [193–195]. For instance, Jinwoo Cheon's group has reported Mn- and Co-doped magnetic nanoparticles, and could generate more thermal energy than conventional $Fe_3O_4$ nanoparticles. Experiments showed that these doped magnetic nanoparticles had enhanced therapeutic efficacy than conventional anticancer drugs, as shown in Fig. 1.27 [193].

Despite the great potential in cancer therapy, the electromagnetic energy-to-heat conversion efficiency is low for currently studied magnetic nanoparticles, which remains a great challenge.

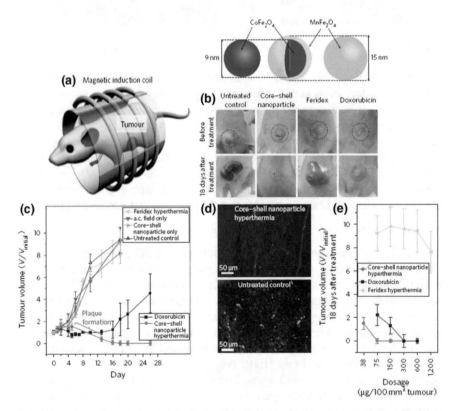

**Fig. 1.27** The use of magnetic nanoparticles for magnetic hyperthermia [193]. (Copyright 2011 Nature)

### 1.3.3.6   Outlook

Multimodal therapy and imaging-guided targeted therapy have become increasingly important in the cancer treatment [196–200]. Indeed, combination of surgery with chemotherapy or radiotherapy is widely used in clinical cancer patients. Each therapeutic method has advantages and also disadvantages. Therefore, integration of multiple therapies is supposed to provide a useful tool for more effective cancer therapy. For example, $O_2$ is required for PDT, while deep tumor tissue is usually hypoxic, and thus limits the therapeutic efficacy of PDT. In comparison, PTT does not need molecular $O_2$. Wang et al. combined PDT and PTT for enhanced treatment of cancer by covalent conjugation of Ce6 on the surface of gold nanoparticles (Fig. 1.28). Upon irradiation with a 671 nm laser, the combined PDT and PTT has demonstrated highest anticancer effect than single PDT or PTT [200].

In case of imaging-guided therapy, the combination of different imaging techniques with different therapies has been investigated, such as MRI-guided ultrasound therapy, MRI-guided PTT, MRI-guided drug delivery, CT imaging-guided PTT, CT-imaging-guided PDT, fluorescence imaging-guided therapy, etc. [201–210]. Yang et al. have developed trimodal MRI, CT and photoacoustic imaging nanoparticles through modification of gold nanorods with carboxyl group-containing

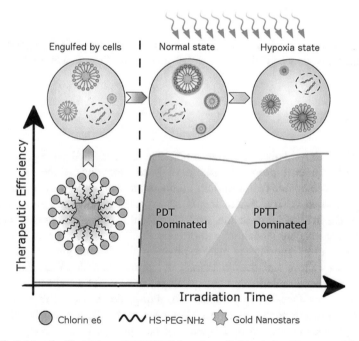

**Fig. 1.28** Schematic illustration of PDT/PTT based on Ce6-loaded gold nanoparticles [201]. (Copyright 2011 Wiley-VCH)

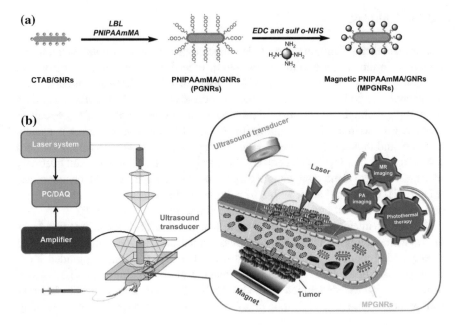

**Fig. 1.29** MRI and photoacoustic imaging-guided photothermal therapy of cancer [210]. (Copyright 2013 Elsevier)

polymer, followed by covalent conjugation to amine-modified $Fe_3O_4$ nanoparticles. Simultaneously, they could be used for PTT therapy, as illustrated in Fig. 1.29 [210].

## 1.4   Significances

Nanomaterials have demonstrated great potential in the biomedical field due to their unique phy-chemical properties and have been emerging as the hot topic of research. By reasonable design and the combination with biology, material science, medicine, and chemistry, nanomaterials are expected to solve the long-term challenges in the field of biomedicine. Multifunctional imaging nanoprobes that combine different imaging modalities can provide solid information for early detection of diseases and monitoring the disease progress. In addition, given the increasing incidence of cancer and the limitations of current therapies including surgery, chemotherapy, and radiotherapy, the development of novel therapeutic approaches and the therapeutic agents, in particular imaging-guided cancer therapy, are highly desirable. Although numerous efforts have been made, the challenges still remain including the synthesis and optimization of nanoprobes, in vivo imaging performance, and therapeutic effects, among others. Aiming at addressing these challenges, in this thesis, we design and synthesize a series of multifunctional nanoprobes for molecular imaging and imaging-guided cancer therapy with enhanced imaging and therapeutic effects. We expect that these multifunctional nanoprobes will provide new tools for effective treatment of advanced human diseases including cancers.

# References

1. Saha K, Agasti SS, Kim C et al (2012) Gold nanoparticles in chemical and biological sensing. Chem Rev 112:2739–2779
2. Singh AK, Senapati D, Neely A et al (2009) Nonlinear optical properties of triangular silver nanomaterials. Chem Pgys Lett 481:94–98
3. http://health.people.com.cn/GB/11713201.html
4. Medintz IL, Uyeda HT, Goldman ER (2005) Mattouss, H. Quantum dot bioconjugates for imaging, labelling and sensing. Nat Mater 4:435–446
5. Drummod DC, Meyer O, Hong K et al (1999) Optimizing liposomes for delivery of chemotherapeutic agents to solid tumors. Pharmacol Rev 51:691–744
6. Chithrani BD, Ghazani AA, Chan WC (2006) Determining the size and shape dependence of gold nanoparticle uptake into mammalian cells. Nano Lett 6:662–668
7. Huang X, Teng X, Chen D et al (2010) The effect of the shape of mesoporous silica nanoparticles on cellular uptake and cell function. Biomaterials 31:438–448
8. Park J, An K, Hwang Y et al (2004) Ultra-large-scale syntheses of monodisperse nanocrystals. Nat Mater 3:891–895
9. Li X, Chen G, Yang L et al (2010) Multifunctional Au-coated $TiO_2$ nanotube arrays as recyclable SERS substrates for multifold organic pollutants detection. Adv Funct Mater 20:2815–2824
10. Wei W, Lu Y, Chen W et al (2011) One-pot synthesis, photoluminescence, and electrocatalytic properties of subnanometer-sized copper clusters. J Am Chem Soc 133:2060–2063
11. Kim S, Fisher B, Eisler HJ (2003) Type-II quantum dots: CdTe/CdSe (core/shell) and CdSe/ZnTe (core/shell) heterostructures. J Am Chem Soc 125:11466–11467
12. Wang F, Liu XG (2009) Chem Soc Rev 38:976–989
13. Hayashi K, Nakamura M, Miki H et al (2012) Near-infrared fluorescent silica/porphyrin hybrid nanorings for in vivo cancer imaging. Adv Funct Mater 22:3539–3546
14. Lummerstorfer T, Hoffmann H (2004) Click chemistry on surfaces: 1,3-dipolar cycloaddition reactions of azide-terminated monolayers on silica. J Phys Chem B 108:3963–3966
15. Wang L, Neoh KG, Kang E-T et al (2011) Multifunctional polyglycerol-grafted $Fe_3O_4@SiO_2$ nanoparticles for targeting ovarian cancer cells. Biomaterials 32:2166–2173
16. Sandiford L, Phinikaridou A, Protti A et al (2013) Bisphosphonate-anchored PEGylation and radiolabeling of superparamagnetic iron oxide: long-circulating nanoparticles for in vivo multimodal (t1 mri-spect) imaging. ACS Nano 7:500–512
17. Hou Y, Qiao R, Fang F et al (2013) $NaGdF_4$ nanoparticle-based molecular probes for magnetic resonance imaging of intraperitoneal tumor xenografts in vivo. ACS Nano 7:330–338
18. Nyk M, Kumar R, Ohulchanskyy TY et al (2008) High contrast in vitro and in vivo photoluminescence bioimaging using near infrared to near infrared up-conversion in $Tm^{3+}$ and $Yb^{3+}$ doped fluoride nanophosphors. Nano Lett 8:3834–3838
19. Boyer JC, Manseau MP, Murray JI et al (2010) Surface modification of upconverting $NaYF_4$ nanoparticles with PEG-phosphate ligands for nir (800 nm) biolabeling within the biological window. Langmuir 26:1157–1164

20. Zhang T, Ge J, Hu Y et al (2007) A general approach for transferring hydrophobic nanocrystals into water. Nano Lett 7:3203–3207
21. Johnson NJJ, Sangeetha NM, Boyer JC et al (2010) Facile ligand-exchange with polyvinylpyrrolidone and subsequent silica coating of hydrophobic upconverting $\beta$-NaYF$_4$:Yb$^{3+}$/Er$^{3+}$ nanoparticles. Nanoscale 2:771–777
22. Cui S, Yin D, Chen Y et al (2013) In vivo targeted deep-tissue photodynamic therapy based on near-infrared light triggered upconversion nanoconstruct. ACS Nano 7:676–688
23. Yu CH, Caiulo N, Lo CCH et al (2006) Synthesis and fabrication of a thin film containing silica-encapsulated face-centered tetragonal FePt nanoparticles. Adv Mater 18:2312–2314
24. Kim J, Kim HS, Lee N et al (2008) Multifunctional uniform nanoparticles composed of a magnetite nanocrystal core and a mesoporous silica shell for magnetic resonance and fluorescence imaging and for drug delivery. Angew Chem Int Ed 47:8438–8441
25. Zhang G, Liu Y, Yuan Q et al (2011) Dual modal in vivo imaging using upconversion luminescence and enhanced computed tomography properties. Nanoscale 3:4365–4371
26. Mahalingam V, Vetrone F, Naccache R et al (2009) Colloidal Tm$^{3+}$/Yb$^{3+}$-Doped LiYF$_4$ nanocrystals: multiple luminescence spanning the uv to nir regions via low-energy excitation. Adv Mater 21:1–4
27. Yi DK, Selvan ST, Lee SS et al (2005) Silica-coated nanocomposites of magnetic nanoparticles and quantum dots. J Am Chem Soc 127:4990–4991
28. Cho NH, Cheong TC, HyunMin J et al (2011) A multifunctional core–shell nanoparticle for dendritic cell-based cancer immunotherapy. Nat Nanotechnol 6:675–682
29. Santra S, Yang H, Holloway PH et al (2005) Synthesis of water-dispersible fluorescent, radio-opaque, and paramagnetic CdS:Mn/ZnS quantum dots: a multifunctional probe for bioimaging. J Am Chem Soc 127:1656–1657
30. Hu KW, Hsu KC, Yeh CS (2010) pH-Dependent biodegradable silica nanotubes derived from Gd(OH)$_3$ nanorods and their potential for oral drug delivery and MR imaging. Biomaterials 31:6843–6848
31. Davis ME, Chen ZG (2008) Shin DM Nanoparticle therapeutics: An emerging treatment modality for cancer. Nat Rev Drug Discovery 7:771–782
32. Peer D, Karp JM, Hong S et al (2007) Nanocarriers as an emerging platform for cancer therapy. Nat Nanotechnol 2:751–760
33. Brannon-Peppas L, Blanchette JO (2004) Nanoparticle and targeted systems for cancer therapy. Adv Drug Delivery Rev 56:1649–1659
34. Park K, Lee S, Kang E et al (2009) New generation of multifunctional nanoparticles for cancer imaging and therapy. Adv Funct Mater 19:1553–1566
35. Bardhan R, Amit joshi S, Halas NJ (2011) Theranostic nanoshells: from probe design to imaging and treatment of cancer. Acc Chem Res 41:936–946
36. Chen D, Feng HB, Li JH (2012) Graphene oxide: preparation, functionalization, and electrochemical applications. Chem Rev 112:6027–6053
37. White KA, Chengelis DA, Gogick KA et al (2009) Near-infrared luminescent lanthanide MOF barcodes. J Am Chem Soc 131:18069–18071
38. Wang L, Yan R, Huo Z et al (2005) Fluorescence resonant energy transfer biosensor based on upconversion-luminescent nanoparticles. Angew Chem Int Ed 44:6054–6057
39. Ai K, Zhang B, Lu L (2009) Europium-based fluorescence nanoparticle sensor for rapid and ultrasensitive detection of an anthrax biomarker. Angew Chem Int Ed 48:304–308
40. Choi Y, Park Y, Kang T et al (2009) Selective and sensitive detection of metal ions by plasmonic resonance energy transfer-based nanospectroscopy. Nat Nanotechnol 4:742–746
41. Ho JA, Chang HC, Su WT (2012) DOPA-mediated reduction allows the facile synthesis of fluorescent gold nanoclusters for use as sensing probes for ferric ions. Anal Chem 84:3246–3253
42. Huang CC, Chen CT, Shiang YC et al (2009) Synthesis of fluorescent carbohydrate-protected au nanodots for detection of concanavalin a and escherichia coli. Anal Chem 81:875–882

43. Wang L, Li Y (2006) Green upconversion nanocrystals for DNA detection. Chem Commun 24:2557–2559

44. Long Y, Jiang D, Zhu X et al (2009) Trace $Hg^{2+}$ analysis via quenching of the fluorescence of a CdS-encapsulated DNA nanocomposite. Anal Chem 81:2652–2657

45. Cao L, Ye J, Tong L et al (2008) A new route to the considerable enhancement of glucose oxidase (GOx) activity: the simple assembly of a complex from CdTe quantum dots and GOx, and its glucose sensing. Chem Eur J 14:9633–9640

46. Deng R, Xie X, Vendrell M et al (2011) Intracellular glutathione detection using $MnO_2$-nanosheet-modified upconversion nanoparticles. J Am Chem Soc 133:20168–20171

47. Wang H, Wang Y, Jin J et al (2008) Gold nanoparticle-based colorimetric and "turn-on" fluorescent probe for mercury(ii) ions in aqueous solution. Anal Chem 80:9021–9028

48. Liu J, Lee JH, Lu Y (2007) Quantum dot encoding of aptamer-linked nanostructures for one-pot simultaneous detection of multiple analytes. Anal Chem 79:4120–4125

49. Teolato P, Rampazzo E, Arduini M et al (2007) Silica nanoparticles for fluorescence sensing of ZnII: exploring the covalent strategy. Chem Eur J 13:2238–2245

50. Bahshi L, Freeman R, Gill R et al (2009) Optical detection of glucose by means of metal nanoparticles or semiconductor quantum dots. Small 5:676–680

51. You CC, Miranda OR, Gider B et al (2007) Detection and identification of proteins using nanoparticle–fluorescent polymer 'chemical nose' sensors. Nat Nanotechnol 2:318–323

52. Liu Q, Peng J, Sun L et al (2011) High-efficiency upconversion luminescent sensing and bioimaging of Hg(II) by chromophoric ruthenium complex-assembled nanophosphors. ACS Nano 5:8040–8048

53. Liu J, Liu Y, Liu Q et al (2011) Iridium(III) complex-coated nanosystem for ratiometric upconversion luminescence bioimaging of cyanide anions. J Am Chem Soc 133:15276–15279

54. Zong C, Ai K, Zhang G et al (2011) Dual-emission fluorescent silica nanoparticle-based probe for ultrasensitive detection of $Cu^{2+}$. Anal Chem 15:3126–3132

55. Si D, Epstein T, Lee YEK et al (2012) Nanoparticle PEBBLE sensors for quantitative nanomolar imaging of intracellular free calcium ions. Anal Chem 84:978–986

56. Kim S, Park JW, Kim D et al (2009) Bioinspired colorimetric detection of calcium(II) ions in serum using calsequestrin-functionalized gold nanoparticles. Angew Chem Int Ed 48:4138–4141

57. Liu JW, Lu Y (2006) Fast colorimetric sensing of adenosine and cocaine based on a general sensor design involving aptamers and nanoparticles. Angew Chem Int Ed 45:90–94

58. Liu JW, Mazumdar D, Lu Y (2006) A simple and sensitive "dipstick" test in serum based on lateral flow separation of aptamer-linked nanostructures. Angew Chem Int Ed 45:7955–7959

59. Li HX, Rothberg L (2004) Colorimetric detection of DNA sequences based on electrostatic interactions with unmodified gold nanoparticles. Proc Natl Acad Sci USA 101:14036–14039

60. Wang ZX, Levy R, Fernig DG et al (2006) Kinase-catalyzed modification of gold nanoparticles: a new approach to colorimetric kinase activity screening. J Am Chem Soc 128:2214–2215

61. Ai KL, Liu YL, Lu LH (2009) Hydrogen-bonding recognition-induced color change of gold nanoparticles for visual detection of melamine in raw milk and infant formula. J Am Chem Soc 131:9496–9497

62. Aslan K, Lakowicz JR, Geddes CD (2004) Nanogold-plasmon-resonance-based glucose sensing. Anal Biochem 330:145–155

63. Aslan K, Lakowicz JR, Geddes CD (2005) Nanogold plasmon resonance-based glucose sensing. 2. wavelength-ratiometric resonance light scattering. Anal Chem 77:2007–2014

64. Baron R, Zayats M, Willner I (2005) Dopamine-, L-DOPA-, adrenaline-, and noradrenaline-induced growth of Au nanoparticles: assays for the detection of neurotransmitters and of tyrosinase activity. Anal Chem 77:1566–1571

65. Radhakumary C, Sreenivasan K (2011) Naked eye detection of glucose in urine using glucose oxidase immobilized gold nanoparticles. Anal Chem 83:2829–2833

66. Liu H, Xu S, He Z et al (2013) Supersandwich cytosensor for selective and ultrasensitive detection of cancer cells using aptamer-DNA concatamer-quantum dots probes. Anal Chem 85:3385–3392

67. Ji J, Yang H, Liu Y et al (2009) $TiO_2$-assisted silver enhanced biosensor for kinase activity profiling. Chem Commun 12:1508–1510

68. Fu Y, Li P, Xie Q et al (2009) One-pot preparation of polymer–enzyme–metallic nanoparticle composite films for high-performance biosensing of glucose and galactose. Adv Funct Mater 19:1784–1791

69. Xi F, Zhao D, Wang X et al (2013) Non-enzymatic detection of hydrogen peroxide using a functionalized three-dimensional graphene electrode. Electrochem Commun 26:81–84

70. Ishikawa FN, Chang HK, Curreli M et al (2009) Label-free, electrical detection of the SARS virus n-protein with nanowire biosensors utilizing antibody mimics as capture probes. ACS Nano 3:1219–1224

71. Roentgen WC (1895) On a new kind of rays. Sitzungsber Phys Med Ges Wurzburg 137:1132–1141

72. 吴晨希, 朱朝晖, 李方 et al (2011) 分子影像: 转化医学的重要工具和主要路径. 生物物理学报 27(4): 327–334

73. Weissleder R (1999) Molecular imaging: exploring the next frontier. Radiology 212(3): 609–614

74. http://www.baike.com/wiki/%E5%88%86%E5%AD%90%E5%BD%B1%E5%83%8F%E5%AD%A6

75. Wu X, Liu H, Liu J (2002) Immunofluorescent labeling of cancer marker Her2 and other cellular targets with semiconductor quantum dots. Nat Biotechnol 21:41–46

76. Efros AL, Rosen M (1997) Random telegraph signal in the photoluminescence intensity of a single quantum dot. Phy Rev Lett 78:1110–1113

77. Wang K, He X, Yang X et al (2013) Functionalized silica nanoparticles: a platform for fluorescence imaging at the cell and small animal levels. Acc Chem Res 46:1367–1376

78. Cai W, Shin DW, Chen K et al (2006) Peptide-labeled near-infrared quantum dots for imaging tumor vasculature in living subjects. Nano Lett 6:669–676

79. Igarashi R, Yoshinari Y, Yokota H et al (2012) Real-time background-free selective imaging of fluorescent nanodiamonds in vivo. Nano Lett 12:5726–5732

80. Wu C, Bull B, Szymanski C et al (2008) Multicolor conjugated polymer dots for biological fluorescence imaging. ACS Nano 2:2415–2423

81. Chan WCW, Nie S (1998) Quantum dot bioconjugates for ultrasensitive nonisotopic detection. Science 281:2016–2018

82. Bruchez M, Moronne M, Gin P et al (1998) Semiconductor nanocrystals as fluorescent biological labels. Science 281:2013–2016

83. http://baike.baidu.com/view/489704.htm

84. Ding C, Zhu A, Tian Y (2013) Functional surface engineering of C-dots for fluorescent biosensing and in vivo bioimaging. Acc Chem Res 47:20–30

85. Eda G, Lin YY, Mattevi C et al (2010) Blue photoluminescence from chemically derived graphene oxide. Adv Mater 22:505–509

86. Rao TUB, Pradeep T (2010) Luminescent $Ag_7$ and $Ag_8$ clusters by interfacial synthesis. Angew Chem Int Ed 49:3925–3929

87. Lu YZ, Chen W (2012) Sub-nanometre sized metal clusters: from synthetic challenges to the unique property discoveries. Chem Soc Rev 41:3594–3623

88. Li W, Zhang Z, Kong B et al (2013) Simple and green synthesis of nitrogen-doped photoluminescent carbonaceous nanospheres for bioimaging. Angew Chem Int Ed 52: 8151–8155

89. Mahalingam V, Vetrone F, Naccache R et al (2009) Colloidal $Tm^{3+}/Yb^{3+}$-Doped $LiYF_4$ nanocrystals: multiple luminescence spanning the uv to nir regions via low-energy excitation. Adv Mater 21:4025–4028

90. Zhang Yu, Zheng F, Yang T et al (2012) Tuning the autophagy-inducing activity of lanthanide-based nanocrystals through specific surface-coating peptides. Nat Mater 11:817–826

91. Li LL, Wu P, Hwang K et al (2013) An exceptionally simple strategy for dna-functionalized upconversion nanoparticles as biocompatible agents for nano-assembly, DNA delivery and imaging. J Am Chem Soc 135:2411–2414

92. Liu Q, Yin B, Yang T et al (2013) A general strategy for biocompatible, high-effective upconversion nanocapsules based on triplet-triplet annihilation. J Am Chem Soc 135:5029–5037

93. Chatterjee DK, Gnanasammandhan MK, Zhang Y (2010) Small upconverting fluorescent nanoparticles for biomedical applications. Small 6:2781–2795

94. Tian G, Gu Z, Zhou L et al (2012) $Mn^{2+}$ dopant-controlled synthesis of $NaYF_4$:Yb/Er upconversion nanoparticles for in vivo imaging and drug delivery. Adv Mater 24:1226–1231

95. Li Z, Zhang Y, Jiang S (2008) Multicolor core/shell-structured upconversion fluorescent nanoparticles. Adv Mater 20:4765–4769

96. Bottrill M, Kwok L, Long NJ (2006) Lanthanides in magnetic resonance imaging. Chem Soc Rev 35:557–571

97. Villaraza AJL, Bumb A, Brechbiel MW (2010) Macromolecules, dendrimers, and nanomaterials in magnetic resonance imaging: the interplay between size, function, and pharmacokinetics. Chem Rev 110:2921–2959

98. Viswanathan S, Kovacs Z, Green KN et al (2010) Alternatives to gadolinium-based metal chelates for magnetic resonance imaging. Chem Rev 110:2960–3018

99. Neves AA, Krishnan AS, Kettunen MI et al (2007) A Paramagnetic nanoprobe to detect tumor cell death using magnetic resonance imaging. Nano Lett 7:1419–1423

100. Schooneveld MM van, Vucic E, Koole R et al (2008) Improved biocompatibility and pharmacokinetics of silica nanoparticles by means of a lipid coating: a multimodality investigation. Nano Lett 8:2517–2525

101. Duncan AK, Klemm PJ, Raymond KN et al (2012) Silica microparticles as a solid support for gadolinium phosphonate magnetic resonance imaging contrast agents. J Am Chem Soc 134:8046–8049

102. Chandrasekharan P, Yong CX, Poh Z et al (2012) Gadolinium chelate with DO3A conjugated 2-(diphenylphosphoryl)-ethyldiphenylphosphonium cation as potential tumor-selective MRI contrast agent. Biomaterials 33:9225–9231

103. Mi P, Cabral H, Kokuryo D et al (2013) Gd-DTPA-loaded polymeremetal complex micelles with high relaxivity for MR cancer imaging. Biomaterials 34:492–500

104. Richard C, Doan BT, Beloeil JC et al (2008) Noncovalent functionalization of carbon nanotubes with amphiphilic $Gd^{3+}$ chelates: toward powerful T1 and T2 MRI contrast agents. Nano Lett 8:232–236

105. Mulder WJM, Strijkers GJ, van Tilborg GAF et al (2006) Lipid-based nanoparticles for contrast-enhanced MRI and molecular imaging. NMR Biomed 19:142–164

106. Cormode DP, Skajaa T, van Schooneveld MM et al (2008) Nanocrystal core high-density lipoproteins: a multimodality contrast agent platform. Nano Lett 8:3715–3723

107. Rieter WJ, Kim JS, Taylor KML et al (2007) Hybrid silica nanoparticles for multimodal imaging. Angew Chem Int Ed 46:3680–3682

108. Bridot JL, Faure AC, Laurent S et al (2007) Gadolinium oxide nanoparticles: multimodal contrast agents for in vivo imaging. J Am Chem Soc 129:5076–5084

109. Kumar R, Nyk M, Ohulchanskyy TY et al (2009) Combined optical and MR bioimaging using rare earth ion doped $NaYF_4$ nanocrystals. Adv Funct Mater 19:853–859

110. Park YI, Kim JH, Lee KT et al (2009) Nonblinking and nonbleaching upconverting nanoparticles as an optical imaging nanoprobe and t1 magnetic resonance imaging contrast agent. Adv Mater 21:4467–4471

111. Chen F, Bu W, Zhang S et al (2013) $Gd^{3+}$-ion-doped upconversion nanoprobes: relaxity mechanism probing and sensitivity optimization. Adv Funct Mater 23:298–307

112. Peng YK, Lai CW, Liu CL et al (2011) A new and facile method to prepare uniform hollow MnO/functionalized mSiO$_2$ core/shell nanocomposites. ACS Nano 5:4177–4187
113. Na HB, Lee JH, An K et al (2007) Development of a T1 contrast agent for magnetic resonance imaging using mno nanoparticles. Angew Chem Int Ed 46:5397–5401
114. Lee YC, Chen DY, Dodd SJ et al (2012) The use of silica coated MnO nanoparticles to control MRI relaxivity in response to specific physiological changes. Biomaterials 33:3560–3567
115. Pan D, Caruthers SD, Senpan A et al (2011) Synthesis of NanoQ, a copper-based contrast agent for high-resolution magnetic resonance imaging characterization of human thrombus. J Am Chem Soc 133:9168–9171
116. Qin J, Laurent S, Jo YS et al (2007) A high-performance magnetic resonance imaging T2 contrast agent. Adv Mater 19:1874–1878
117. Lee N, Choi Y, Lee Y et al (2012) Water-dispersible ferrimagnetic iron oxide nanocubes with extremely high r2 relaxivity for highly sensitive in vivo MRI of tumors. Nano Lett 12:3127–3131
118. Kim BH, Lee N, Kim H et al (2011) Large-scale synthesis of uniform and extremely small-sized iron oxide nanoparticles for high-resolution T1 magnetic resonance imaging contrast agents. J Am Chem Soc 133:12624–12631
119. Jakhmola A, Anton N, Vandamme TF (2012) Inorganic nanoparticles based contrast agents for X-ray computed tomography. Adv Healthcare Mater 1:413–431
120. Haller C, Hizoh I (2004) In vitro cytotoxic effects of iodinated contrast media on a renal tubular cell line. Invest Radiol 39:149–154
121. Lusic H, Grinstaff MW (2013) X-ray-computed tomography contrast agents. Chem Rev 113:1641–1666
122. Kim D, Park S, Lee JH et al (2007) Antibiofouling polymer-coated gold nanoparticles as a contrast agent for in vivo X-ray computed tomography imaging. J Am Chem Soc 129:7661–7665
123. Eck W, Nicholson AI, Zentgraf H et al (2010) Anti-CD4-targeted gold nanoparticles induce specific contrast enhancement of peripheral lymph nodes in x-ray computed tomography of live mice. Nano Lett 10:2318–2322
124. Sun IC, Eun DK, Na JH et al (2009) Heparin-coated gold nanoparticles for liver-specific CT imaging. Chem Eur J 15:13341–13347
125. Chou SW, Shau YH, Wu PC et al (2010) In vitro and in vivo studies of FePt nanoparticles for dual modal CT/MRI molecular imaging. J Am Chem Soc 132:13270–13278
126. Rabin O, Perez JM, Grimm J et al (2006) An X-ray computed tomography imaging agent based on long-circulating bismuth sulphide nanoparticles. Nat Mater 5:118–122
127. Oh MH, Lee N, Kim H et al (2011) Large-scale synthesis of bioinert tantalum oxide nanoparticles for x-ray computed tomography imaging and bimodal image-guided sentinel lymph node mapping. J Am Chem Soc 133:5508–5515
128. Schutt EG, Klein DH, Mattrey RM et al (2003) Injectable microbubbles as contrast agents for diagnostic ultrasound imaging: the key role of perfluorochemicals. Angew Chem Int Ed 42:3218–3235
129. Lin PL, Eckersley RJ, Hall EAH (2009) Ultrabubble: a laminated ultrasound contrast agent with narrow size range. Adv Mater 21:3949–3952
130. Lopez RD, Tsapis N, Libong D et al (2009) Phospholipid decoration of microcapsules containing perfluorooctyl bromide used as ultrasound contrast agents. Biomaterials 30:1462–1472
131. Wang X, Chen H, Chen Y et al (2012) Perfluorohexane-encapsulated mesoporous silica nanocapsules as enhancement agents for highly efficient high intensity focused ultrasound (HIFU). Adv Mater 24:789–791
132. Ku G, Zhou M, Song S et al (2012) Copper sulfide nanoparticles as a new class of photoacoustic contrast agent for deep tissue imaging at 1064 nm. ACS Nano 6:7489–7496
133. Homan KA, Souza M, Truby R et al (2012) Silver nanoplate contrast agents for in vivo molecular photoacoustic imaging. ACS Nano 6:641–650

134. Liu Z, Cai W, He L et al (2007) In vivo biodistribution and highly efficient tumour targeting of carbon nanotubes in mice. Nat Nanotechnol 2:47–52
135. Huynh E, Lovell JF, Helfield B et al (2012) Porphyrin shell microbubbles with intrinsic ultrasound and pho-toacoustic properties. J Am Chem Soc 134:16464–16467
136. Zerda ADL, Zavaleta C, Keren S et al (2008) Carbon nanotubes as photoacoustic molecular imaging agents in living mice. Nat Nanotechnol 3:557–562
137. Hong H, Yang K, Zhang Y et al (2012) In vivo targeting and imaging of tumor vasculature with radiolabeled, antibody-conjugated nanographene. ACS Nano 6:2361–2370
138. Wang Y, Liu Y, Luehmann H et al (2012) Evaluating the pharmacokinetics and in vivo cancer targeting capability of Au nanocages by positron emission tomography imaging. ACS Nano 6:5880–5888
139. Wong RM, Gilbert DA, Liu K et al (2012) Rapid Size-controlled synthesis of dextran-coated, $^{64}$Cu-doped iron oxide nanoparticles. ACS Nano 6:3461–3467
140. Mahmoudi M, Serpooshan V, Laurent S (2011) Engineered nanoparticles for biomolecular imaging. Nanoscale 3:3007–3026
141. Burns AA, Vider J, Ow H et al (2009) Fluorescent silica nanoparticles with efficient urinary excretion for nanomedicine. Nano Lett 9:442–448
142. Lee N, Cho HR, Oh MH et al (2012) Multifunctional $Fe_3O_4$/TaOx core/shell nanoparticles for simultaneous magnetic resonance imaging and x-ray computed tomography. J Am Chem Soc 134:10309–10312
143. Banerjee SR, Pullambhatla M, Byun Y et al (2011) Sequential SPECT and optical imaging of experimental models of prostate cancer with a dual modality inhibitor of the prostate-specific membrane antigen. Angew Chem Int Ed 50:9167–9170
144. Xia A, Chen M, Gao Y et al (2012) $Gd^{3+}$ complex-modified $NaLuF_4$-based upconversion nanophosphors for trimodality imaging of NIR-to-NIR upconversion luminescence, X-Ray computed tomography and magnetic resonance. Biomaterials 33:5394–5405
145. Bruns OT, Ittrich H, Peldschus K et al (2009) Real-time magnetic resonance imaging and quantification of lipoprotein metabolism in vivo using nanocrystals. Nat Nanotechnol 4:193–201
146. Harrisson S, Nicolas J, Maksimenko A et al (2012) Nanoparticles with in vivo anticancer activity from polymer prodrug amphiphiles prepared by living radical polymerization. Angew Chem Int Ed 51:1678–1682
147. Peng F, Su Y, Wei X et al (2012) Silicon-nanowire-based nanocarriers with ultrahigh drug-loading capacity for in vitro and in vivo cancer therapy. Angew Chem Int Ed 51:1457–1461
148. Qi C, Zhu YJ, Zhao XY et al (2012) Highly stable amorphous calcium phosphate porous nanospheres: microwave-assisted rapid synthesis using atp as phosphorus source and stabilizer, and their application in anticancer drug delivery. Chem Eur J 19:981–987
149. Xing L, Zheng H, Cao Y et al (2012) Coordination polymer coated mesoporous silica nanoparticles for ph-responsive drug release. Adv Mater 24:6433–6437
150. Zhang ZY, Xu YD, Ma YY et al (2013) Biodegradable ZnO@polymer core–shell nanocarriers: pH-triggered release of doxorubicin in vitro. Angew Chem Int Ed 52:4127–4131
151. Zhang J, Yuan ZF, Wang Y et al (2013) Multifunctional envelope-type mesoporous silica nanoparticles for tumor-triggered targeting drug delivery. J Am Chem Soc 135:5068–5073
152. Fang W, Yang J, Gong J et al (2011) Photo-and pH-triggered release of anticancer drugs from mesoporous silica-coated Pd@Ag nanoparticles. Adv Funct Mater 22:842–848
153. Zhang X, Yang P, Dai Y et al (2013) Multifunctional up-converting nanocomposites with smart polymer brushes gated mesopores for cell imaging and thermo/pH dual-responsive drug controlled release. Adv Funct Mater 23:4067–4078
154. Huang P, Lin J, Wang S et al (2013) Photosensitizer-conjugated silica-coated gold nanoclusters for fluorescence imaging-guided photodynamic therapy. Biomaterials 34: 4643–4653

155. Zhu Z, Tang Z, Phillips JA et al (2008) Regulation of singlet oxygen generation using single-walled carbon nanotubes. J Am Chem Soc 130:10856–10857

156. Brasch M, Escosura A, Ma Y et al (2011) Encapsulation of phthalocyanine supramolecular stacks into virus-like particles. J Am Chem Soc 133:6878–6881

157. Cheng Y, Samia AC, Meyers JD et al (2008) Highly efficient drug delivery with gold nanoparticle vectors for in vivo photodynamic therapy of cancer. J Am Chem Soc 130:10643–10647

158. Ratanatawanate C, Chyao A, Jr KJB (2011) S-Nitrosocysteine-decorated PbS QDs/TiO$_2$ nanotubes for enhanced production of singlet oxygen. J Am Chem Soc 133:3492–3497

159. Kim S, Ohulchanskyy TY, Pudavar HE et al (2007) Organically modified silica nanoparticles Co-encapsulating photosensitizing drug and aggregation-enhanced two-photon absorbing fluorescent dye aggregates for two-photon photodynamic therapy. J Am Chem Soc 129:2669–2675

160. Liang X, Li X, Yue X et al (2011) Conjugation of porphyrin to nanohybrid cerasomes for photodynamic diagnosis and therapy of cancer. Angew Chem Int Ed 50:11622–11627

161. Park YI, Kim HM, Kim JH et al (2012) Theranostic Probe Based on Lanthanide-doped nanoparticles for simultaneous in vivo dual-modal imaging and photodynamic therapy. Adv Mater 24:5755–5761

162. Liu K, Liu X, Zeng Q et al (2012) Covalently assembled NIR nanoplatform for simultaneous fluorescence imaging and photodynamic therapy of cancer cells. ACS Nano 6:4054–4062

163. Wang C, Cheng L, Liu Y et al (2013) Imaging-guided pH-sensitive photodynamic therapy using charge reversible upconversion nanoparticles under near-infrared light. Adv Funct Mater 23:3077–3086

164. Wang C, Tao H, Cheng L et al (2011) Near-infrared light induced in vivo photodynamic therapy of cancer based on upconversion nanoparticles. Biomaterials 32:6145–6154

165. Qian HS, Guo HC, Ho PCL et al (2009) Mesoporous-silica-coated up-conversion fluorescent nanoparticles for photodynamic therapy. Small 5:2285–2290

166. Choi WI, Kim JY, Kang C et al (2011) Tumor regression in vivo by photothermal therapy based on gold-nanorod-loaded, functional nanocarriers. ACS Nano 5:1995–2003

167. Xiao Z, Ji C, Shi J et al (2012) DNA self-assembly of targeted near-infrared-responsive gold nanoparticles for cancer thermo-chemotherapy. Angew Chem Int Ed 51:11853–11857

168. Lee SM, Kim HJ, Ha YJ et al (2013) Targeted chemo-photothermal treatments of rheumatoid arthritis using gold half-shell multifunctional nanoparticles. ACS Nano 7:50–57

169. Gao L, Fei J, Zhao J et al (2012) Hypocrellin loaded gold nanocages with high two-photon efficiency for the photothermal/photodynamic cancer therapy in vitro. ACS Nano 6: 8030–8040

170. Yang J, Lee J, Kang J et al (2009) Smart drug-loaded polymer gold nanoshells for systemic and localized therapy of human epithelial cancer. Adv Mater 21:4339–4342

171. Zhang W, Guo Z, Huang D et al (2011) Synergistic effect of chemo-photothermal therapy using PEGylated graphene oxide. Biomaterials 32:8555–8561

172. Akhavan O, Ghaderi E (2013) Graphene nanomesh promises extremely efficient in vivo photothermal therapy. Small 9:3593–3601

173. Kam NWS, O'Connell M, Wisdom JA et al (2005) Carbon nanotubes as multifunctional biological transporters and near-infrared agents for selective cancer cell destruction. Proc Natl Acad Sci U S A 102:11600–11605

174. Wang X, Wang C, Cheng L et al (2012) Noble metal coated single-walled carbon nanotubes for applications in surface enhanced raman scattering imaging and photothermal therapy. J Am Chem Soc 134:7414–7422

175. Markovic ZM, Harhaji-Trajkovic LM, Todorovic-Markovic BM et al (2011) In vitro comparison of the photothermal anticancer activity of graphene nanoparticles and carbon nanotubes. Biomaterials 32:1121–1129

176. Liu X, Tao H, Yang K et al (2011) Optimization of surface chemistry on single-walled carbon nanotubes for in vivo photothermal ablation of tumors. Biomaterials 32:144–151

177. Tian Q, Jiang F, Zou R et al (2011) Hydrophilic $Cu_9S_5$ nanocrystals: a photothermal agent with a 25.7% heat conversion efficiency for photothermal ablation of cancer cells in vivo. ACS Nano 5:9761–9771

178. Song G, Wang Q, Wang Y et al (2013) A low-toxic multifunctional nanoplatform based on $Cu_9S_5$ @mSiO$_2$ core-shell nanocomposites: combining photothermal- and chemotherapies with infrared thermal imaging for cancer treatment. Adv Funct Mater 23:4281–4292

179. Dong K, Liu Z, Li Z et al (2013) Hydrophobic Anticancer drug delivery by a 980 nm laser-driven photothermal vehicle for efficient synergistic therapy of cancer cells in vivo. Adv Mater 25:4452–4458

180. Tian Q, Tang M, Sun Y et al (2011) Hydrophilic flower-like CuS superstructures as an efficient 980 nm laser-driven photothermal agent for ablation of cancer cells. Adv Mater 23:3542–3547

181. Hessel CM, Pattani VP, Rasch M et al (2011) Copper selenide nanocrystals for photothermal therapy. Nano Lett 11:2560–2566

182. Li W, Zamani R, Gil PR et al (2013) CuTe Nanocrystals: Shape and size control, plasmonic properties, and use as SERS probes and photothermal agents. J Am Chem Soc 135:7098–7101

183. Chu M, Pan X, Zhang D et al (2012) The therapeutic efficacy of CdTe and CdSe quantum dots for photothermal cancer therapy. Biomaterials 33:7071–7083

184. Chou SS, Kaehr B, Kim J et al (2013) Chemically exfoliated $MoS_2$ as near-infrared photothermal agents. Angew Chem Int Ed 52:4160–4164

185. Chen Z, Wang Q, Wang H et al (2013) Ultrathin PEGylated $W_{18}O_{49}$ nanowires as a new 980 nm-laser-driven photothermal agent for efficient ablation of cancer cells in vivo. Adv Mater 25:2095–2100

186. Wang S, Kim G, Lee YEK et al (2012) Multifunctional biodegradable polyacrylamide nanocarriers for cancer theranostics—a "see and treat" strategy. ACS Nano 6:6843–6851

187. Cheng L, Yang K, Chen Q et al (2012) Organic Stealth nanoparticles for highly effective in vivo near-infrared photothermal therapy of cancer. ACS Nano 6:5605–5613

188. Zha Z, Yue X, Ren Q et al (2012) Uniform polypyrrole nanoparticles with high photothermal conversion efficiency for photothermal ablation of cancer cells. Adv Mater 25:777–782

189. Yang K, Xu H, Cheng L et al (2012) In vitro and in vivo near-infrared photothermal therapy of cancer using polypyrrole organic nanoparticles. Adv Mater 24:5586–5592

190. Yang J, Choi J, Bang D et al (2011) Convertible organic nanoparticles for near-infrared photothermal ablation of cancer cells. Angew Chem Int Ed 50:441–444

191. Lovell JF, Jin CS, Huynh E et al (2011) Porphysome nanovesicles generated by porphyrin bilayers for use as multimodal biophotonic contrast agents. Nat Mater 10:324–332

192. Lovell JF, Jin CS, Huynh E et al (2012) Enzymatic regioselection for the synthesis and biodegradation of porphysome nanovesicles. Angew Chem Int Ed 51:2429–2433

193. Lee JH, Jang JT, Choi J et al (2011) Exchange-coupled magnetic nanoparticles for efficient heat induction. Nat Nanotechnol 6:418–422

194. Bae KH, Park M, Do MJ et al (2012) Chitosan oligosaccharide-stabilized ferrimagnetic iron oxide nanocubes for magnetically modulated cancer hyperthermia. ACS Nano 6:5266–5273

195. Sadhukha T, Wiedmann TS, Panyam J (2013) Inhalable magnetic nanoparticles for targeted hyperthermia in lung cancer therapy. Biomaterials 34:5163–5171

196. Zheng M, Yue C, Ma Y et al (2013) Single-step assembly of DOX/ICG loaded lipid-polymer nanoparticles for highly effective chemo-photothermal combination therapy. ACS Nano 7:2056–2067

197. Sheng Z, Song L, Zheng J et al (2013) Protein-assisted fabrication of nano-reduced graphene oxide for combined in vivo photoacoustic imaging and photothermal therapy. Biomaterials 34:5236–5243

198. Lee JH, Chen KJ, Noh SH et al (2013) On-demand drug release system for in vivo cancer treatment through self-assembled magnetic nanoparticles. Angew Chem Int Ed 52:4384–4388

199. Tong R, Hemmati HD, Langer R et al (2012) Photoswitchable nanoparticles for triggered tissue penetration and drug delivery. J Am Chem Soc 134:8848–8855
200. Wang S, Huang P, Nie L et al (2013) Single continuous wave laser induced photodynamic/plasmonic photothermal therapy using photosensitizer-functionalized gold nanostars. Adv Mater 25:3055–3061
201. Chen Y, Chen H, Sun Y et al (2011) Multifunctional Mesoporous Composite nanocapsules for highly efficient MRI-guided high-intensity focused ultrasound cancer surgery. Angew Chem Int Ed 50:12505–12509
202. Cheng L, Yang K, Li Y et al (2011) Facile preparation of multifunctional upconversion nanoprobes for multimodal imaging and dual-targeted photothermal therapy. Angew Chem Int Ed 50:7385–7390
203. Hu SH, Chen YW, Hung WT et al (2012) Quantum-dot-tagged reduced graphene oxide nanocomposites for bright fluorescence bioimaging and photothermal therapy monitored in situ. Adv Mater 24:1748–1754
204. You JO, Guo P, Auguste DT (2013) A drug-delivery vehicle combining the targeting and thermal ablation of HER2 + breast-cancer cells with triggered drug release. Angew Chem Int Ed 52:4141–4146
205. Yang J, Lee CH, Ko HJ et al (2007) Multifunctional magneto-polymeric nanohybrids for targeted detection and synergistic therapeutic effects on breast cancer. Angew Chem Int Ed 46:8836–8839
206. Huh YM, Lee ES, Lee JH et al (2007) Hybrid nanoparticles for magnetic resonance imaging of target-specific viral gene delivery. Adv Mater 19:3109–3112
207. Bhirde AA, Patel V, Gavard J et al (2009) Targeted killing of cancer cells in vivo and in vitro with EGF-directed carbon nanotube-based drug delivery. ACS Nano 3:307–316
208. Yoo D, Jeong H, Preihs C (2012) Double-effector nanoparticles: a synergistic approach to apoptotic hyperthermia. Angew Chem Int Ed 51:12482–12485
209. Chen H, Li B, Ren X et al (2012) Multifunctional near-infrared-emitting nano-conjugates based on gold clusters for tumor imaging and therapy. Biomaterials 33:8461–8476
210. Yang HW, Liu HL, Li ML et al (2013) Magnetic gold-nanorod/PNIPAAmMA nanoparticles for dual magnetic resonance and photoacoustic imaging and targeted photothermal therapy. Biomaterials 34:5651–5660

# Chapter 2
# Application of Gold-Nanocluster-Based Fluorescent Sensors for Highly Sensitive and Selective Detection of Cyanide in Water

**Abstract** Cyanide is a highly toxic substance and can invade the human body through many routes. Nevertheless, cyanide is still used in many practical fields. Thus, there is an urgent need to develop highly sensitive and selective sensing systems for the detection of cyanide in the environment, especially in water and biological samples. To overcome the problems of currently studied cyanide sensors such as poor water solubility, poor selectivity, and complex preparation procedures, we develop an innovative gold nanocluster-based fluorescent sensor for cyanide in aqueous solutions. Owing to the unique Elsner reaction between cyanide and the gold atoms of gold nanoclusters, this sensor shows high sensitivity and strong tolerance to other interferences. More impressively, this sensor can be directly used for the detection of cyanide in aqueous solution with excellent recoveries. Thus, this gold-nanocluster-based sensor may provide an effective new tool for highly sensitive and selective detection of cyanide in biological samples.

## 2.1 Introduction

In USA, approximately 5000–10,000 people each year die of inhalation of the smoke emitted from burning of plastics, woolens, or other nitrogen-containing substances [1, 2]. Cyanide is the most important prime culprit. Cyanide is one of the most-concerning anions in the environment, the toxicity of which is known because of its binding to a3 cytochromes and inhibition of the electron transport chain in mitochondria [3]. Different from toxic metal ions, which induce some diseases mainly by accumulating in the human body, cyanide can directly lead to the death of human beings as well as aquatic life in several minutes, even at a low concentration, by depressing the central nervous system. In addition, cyanides can invade human body through various routes including inhalation, skin contact, or injection. Therefore, cyanides are widely used for murder by criminals. Despite its high toxicity, cyanide is still widely used in electroplating, gold mining and other fields, and approximately 14,00,000 t of toxic cyanide is produced per year [4]. Accidental

© Springer Nature Singapore Pte Ltd. 2018
Y. Liu, *Multifunctional Nanoprobes*, Springer Theses,
DOI 10.1007/978-981-10-6168-4_2

cyanide release can thus result in serious contamination of the groundwater and even drinking water, leading to human death.

Conventional detection of cyanide relies on several methods including electrometric, chromatographic, titrimetric, and voltammetric techniques. However, these methods are generally time-consuming, have complicated procedures, and request significant special skills. To overcome these issues, colorimetric strategies, a cost-effective and naked-eye detection method have been used for cyanide sensing, based on the strong nucleophilicity or high binding affinity of cyanide with the sensors [5–7]. Nevertheless, these methods also showed limitations, such as complicated organic synthesis and a high detection limit. Fluorescence, in contrast, has been emerging as a more-powerful optical technique for the detection of low concentration of analytes, owing to its simple, inexpensive, and rapid implementation. As a result, varieties of fluorescent cyanide chemosensors have been developed. Although these optical chemosensors have made great contributions in cyanide sensing, these strategies also pose limitations, such as environmentally harmful systems, water-insolubility, poor photostability, a high detection limit or easy interference from other anions (especially by $F^-$, $Ac^-$) [8–15]. For the detection of cyanide in the biosamples, it requires the sensors to have good water solubility, high sensitivity, and selectivity. Moreover, facile synthesis is also needed for practical applications.

To achieve this goal, in this chapter, we describe a novel gold-nanocluster (Au NC)-based fluorescent sensor for the recognition and determination of cyanide in aqueous solutions, based on the fluorescence quenching effect of Au NCs by cyanide. Notably, several advantages of this method make it especially attractive for the detection of cyanide in the real samples: (i) the sensor was synthesized with a facile and environmentally friendly synthetic route, minimizing the cost and avoiding the use of toxic organic reagents; (ii) the fluorescence of Au NCs is highly size-dependent, and thus highly sensitive to cyanide etching; (iii), the fluorescent, nontoxic metal nanoclusters can be applied directly as a sensing moiety, which enables the facile detection of cyanide; and (iv) gold is chemically inert, and few anions can react with it, except cyanide, enabling a high selectivity toward cyanide. Importantly, this sensing system can work directly in aqueous solution, thus enabling monitoring of practical samples, such as biofluids.

## 2.2 Experimental Section

### 2.2.1 Materials

BSA was purchased from Beijing Dingguo Chemical and Biotechnology Co. Ltd. $HAuCl_4 \cdot 3H_2O$ was obtained from Alfa. All other reagents and solvents are of analytical grade and used without further purification.

### 2.2.2  *Preparation of the BSA-Stabilized Au NCs*

The BSA-stabilized Au NCs were synthesized according to a method reported previously [16]. Briefly, HAuCl$_4$ solution (5 mL, 10 mM) was mixed with BSA solution (5 mL, 50 mg/mL) under vigorous stirring at 37 °C. Two minutes later, NaOH solution (0.5 mL, 1 M) was added, and the reaction was allowed to proceed for 12 h under vigorous stirring at 37 °C.

### 2.2.3  *Fluorescent Detection of Cyanide*

The KCN stock solution was prepared by dissolving KCN (0.12 M) in a NaOH–NaHCO$_3$ buffer (0.01 M) at pH 12.0. CN$^-$ solutions with various concentrations were obtained by serial dilution of the stock solution with the buffer. For CN$^-$ detection, the CN$^-$ solutions with different concentrations were added into the Au NCs solution in the NaOH–NaHCO$_3$ buffer at pH 12.0; the mixture solution was equilibrated for 20 min before measurements.

### 2.2.4  *Selectivity Measurement*

The selectivity of this sensing system for cyanide was evaluated by monitoring the fluorescence response of Au NCs to other common anions using the following salts: Na$_2$S, KSCN, NaBr, Na$_2$CO$_3$, NaCl, NaNO$_3$, Na$_2$SO$_4$, NaN$_3$, KCN, EDTA disodium salt, CH$_3$COONa (NaAc), Na$_3$PO$_4$, NaNO$_2$, NaIO$_3$, sodium citrate, Na$_2$C$_2$O$_4$, NaF, and NaI. The above salt solutions were respectively mixed with the Au NCs solutions in a NaOH–NaHCO$_3$ buffer (0.01 M, pH of 12.0) and equilibrated for 20 min before measurements.

For the detection of selectivity of this sensing system over cations, Fe$^{3+}$, Zn$^{2+}$, Ni$^{2+}$, Cd$^{2+}$, Hg$^{2+}$, Pb$^{2+}$, Cu$^{2+}$, Co$^{2+}$, Ag$^+$, K$^+$, Na$^+$, Li$^+$, Mg$^{2+}$, Ca$^{2+}$, Ba$^{2+}$, Al$^{3+}$ were used.

### 2.2.5  *Detection of Cyanide in Real Samples*

The tap-water and groundwater samples were used without any pretreatment, while the lake-water and pond-water samples were filtered twice using a syringe filter (pore diameter 0.02 mm) to remove oil and other organic impurities prior to analysis. Water samples spiked with various concentrations of cyanide were added to the Au NCs, and the fluorescence detections were performed within 20 min.

## 2.2.6 Characterization

The fluorescence spectra were obtained using a Perkin–Elmer LS 55 luminescence spectrometer. XPS was performed using a VG ESCALAB MKII spectrometer. The XPSPEAK software (Version 4.1) was used to deconvolute the narrow-scan XPS spectra of the Au 4f of the Au NCs, using adventitious carbon to calibrate the C1s binding energy (284.5 eV). The elemental analysis was performed using an ELAN 9000/DRC ICP-MS system. The molecular weights of the Au NCs in the absence and presence of cyanide were determined using matrix-assisted laser desorption ionization mass spectrometry (MALDI-MS).

## 2.3 Results and Discussion

### 2.3.1 Mechanism for Cyanide Detection with Au NCs

Au NCs are composed of several to tens of atoms with the size less than 1 nm, which is close to the Fermi wavelength of an electron. The spatial confinement of free electrons in Au NCs results in discrete and size-tunable electronic transitions, thus offering molecular like properties including fluorescence. Interestingly, the fluorescence of Au NCs is size-dependent, which we hypothesized that these fluorescent nanoclusters could be used for the detection of cyanide.

First, we prepared BSA-stabilized Au NCs. The aqueous solution of the Au NCs was deep brown in color and exhibited bright-red fluorescence under irradiation with the 365 nm UV light. To investigate whether these fluorescent nanoclusters could be used for the detection of cyanide, 5 mM cyanide was added in the aqueous solution of the Au NCs. We found that the original deep-brown color of the Au NCs became colorless, and very-weak blue fluorescence was observed under irradiation with a 365 nm UV light (Fig. 2.1). The corresponding characteristic fluorescence emission of the Au NCs at 640 nm was also disappeared, suggesting the reaction between Au NCs and cyanide (Fig. 2.2).

Given these phenomena, a series of experiments were carried out to reveal the mechanisms behind the reaction of Au NCs and cyanide. First, X-ray photoelectron spectroscopy (XPS) was used to investigate the oxidation states of the Au NCs with vs without addition of cyanide. For the Au NCs, the Au 4f7/2 spectrum could be deconvoluted into two peaks centered at 83.5 eV and 85.1 eV, which are corresponding to the binding energies of Au(0) and Au(I), respectively; whereas, upon the addition of 5 mM cyanide, only an enhanced Au(I) peak was observed at 85.1 eV (Fig. 2.3), indicating the oxidation of Au atoms in the Au NCs by the cyanide.

In previous reports, matrix-assisted laser desorption ionization mass spectrometry (MALDI-MS) was often used to determine the number of Au atoms in Au NCs [17]. Thus, the number of Au atoms in our Au NCs before versus after reaction with

**(a)**                                   **(b)**

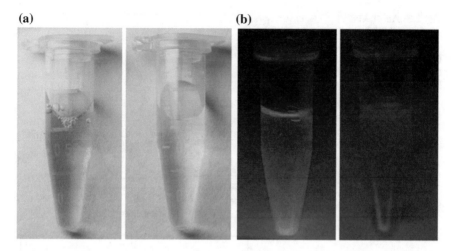

**Fig. 2.1** **a** The digital pictures of BSA-stabilized Au NCs in the absence versus presence of 5 mM cyanide. **b** The corresponding fluorescent pictures of BSA-stabilized Au NCs in the absence versus presence of 5 mM cyanide under irradiation with the 365 UV light. (Copyright 2010 Wiley-VCH)

**Fig. 2.2** Excitation and emission spectra of BSA-stabilized Au NCs in the absence versus presence of 5 mM cyanide. (Copyright 2010 Wiley-VCH)

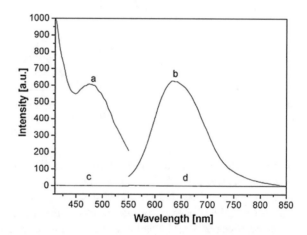

cyanide was determined by MALDI-MS to further confirm the oxidation reaction between Au NCs and cyanide. Prior to detection, a dialysis procedure was performed to purify the Au NCs, followed by freeze-drying. As shown in Fig. 2.4, BSA-stabilized Au NCs exhibited a characteristic peak at 70 kDa. Given the molecular weight of BSA (67 kDa), the number of gold atoms in Au NCs is calculated to be 25. We noticed that the peak at 70 kDa, characteristic of the BSA-stabilized Au NCs, disappeared upon addition of 5 mM cyanide. We reasoned that when the BSA-stabilized Au NCs were dialyzed against water using a 10 kDa cut-off dialysis bag, they could not leach into the filtrate due to their large size. In contrast, cyanide reacted with the BSA-stabilized Au NCs, resulting in the release of Au atoms from the BSA-stabilized Au NCs, thus allowing the resulting Au atoms

**Fig. 2.3** Au 4f XPS spectrum of BSA-stabilized Au NCs in the absence (**a**) versus presence (**b**) of 5 mM cyanide. (Copyright 2010 Wiley-VCH)

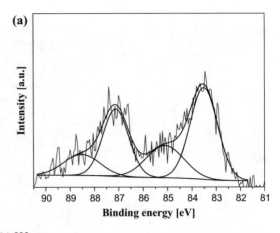

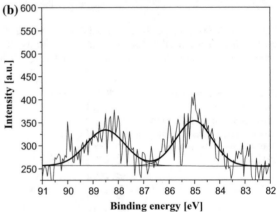

to be filtrated through the cut-off dialysis bag. To further confirm our hypothesis, ICP analysis was carried out to detect the concentration of gold atoms in the filtrate of BSA-stabilized Au NCs in the absence vs presence of 5 mM cyanide. Results showed that the concentration of Au ions in the filtrate of BSA-stabilized Au NCs in the presence of 5 mM cyanide was 648.7 mg/mL, strongly indicating the etching reaction between BSA-stabilized Au NCs and cyanide.

Previous studies have proved that cyanide reacts with Au to generate a very stable $Au(CN)^{2-}$ complex through strong covalent bonding, which is referred to as the Elsner reaction [18, 19]:

$$4Au + 8CN^- + 2H_2O + O_24Au(CN)^{2-} + 4OH^- \tag{2.1}$$

According to Eq. 2.1, cyanide can etch the Au NCs, thus leading to the fluorescence quenching of the Au NCs at 640 nm. The reaction mechanism is demonstrated in Fig. 2.5. Notably, the very-weak blue fluorescence, as aforementioned, is characteristic emission of the aromatic side groups in the amino acid residues of the BSA (tryptophan, tyrosine and phenylalanine).

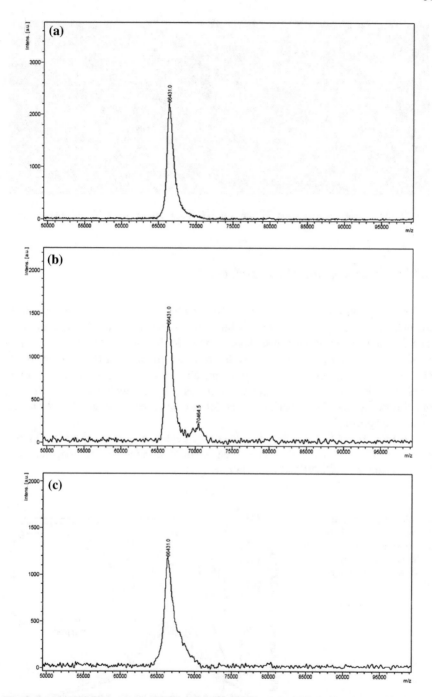

**Fig. 2.4** MALDI-MS spectrum of BSA (**a**), BSA-stabilized Au NCs in the absence (**b**) versus presence (**c**) of 5 mM cyanide. (Copyright 2010 Wiley-VCH)

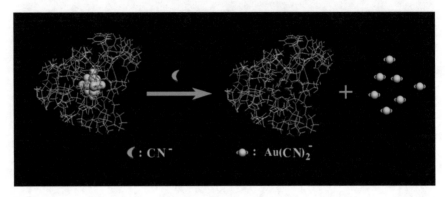

**Fig. 2.5** Schematic representation of the Au NC-based sensor for CN⁻. (Copyright 2010 Wiley-VCH)

## 2.3.2  Optimization of the Detection

Before using our fluorescent sensor for the detection of cyanide, we first optimized the detection time, given the slow etching process between Au NCs and cyanide. The time-dependent fluorescence change upon addition of 50 μM cyanide was first monitored. As shown in Fig. 2.6, the fluorescence intensity at 640 nm was quenched by approximately 47% within 1 min and, remained constant after 20 min etching, indicating that the etching reaction was complete within 20 min. Thus, all the following tests were performed at 20 min after addition of cyanide unless otherwise specified.

Another crucial factor is the pH value of the detection system. Cyanide is known to be a week acid, and a thus may exist in forms in aqueous solution, including hydrocyanic acid (HCN) and cyanide ions (CN⁻):

**Fig. 2.6** Time-dependent fluorescence response of BSA-stabilized Au NCs upon addition of 5 mM cyanide. (Copyright 2010 Wiley-VCH)

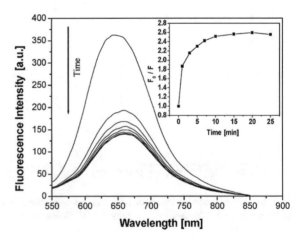

**Fig. 2.7** Fluorescence response of BSA-stabilized Au NCs upon addition of 5 mM cyanide under different pH values. The insert is the relative fluorescence changes of BSA-stabilized Au NCs upon addition of 5 mM cyanide under different pH values. (Copyright 2010 Wiley-VCH)

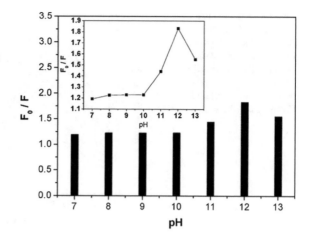

$$CN^- + H_2O \rightleftharpoons HCN + OH^- \ (pKa = 9.36) \qquad (2.2)$$

Nevertheless, only $CN^-$ can etch Au NCs. According to the Eq. 2.2, when pH is decreased, the equilibrium will move to right, leading to decreased $CN^-$. In contrast, when pH is increased, this equilibrium will move to left, thus resulting in increased $CN^-$. To this end, the pH effect on the sensing system was examined. Figure 2.7 is the response of Au NCs to cyanide under different pH values, and the Au NCs have demonstrated highest sensitivity against cyanide at pH 12.

### 2.3.3   Sensitivity Analysis

Given the optimized reaction time and pH, we proceeded to test the sensitivity of the Au NCs against cyanide. Different concentrations of cyanide were respectively added into the aqueous solution of Au NCs and the fluorescence intensity of Au NCs was recorded. As seen in Fig. 2.8, a dose-dependent decrease in the fluorescence intensity of Au NCs was observed with increasing concentration of cyanide. The quenching efficiency was fitted to the Stern–Volmer equation [20]:

$$F_0/F = 1 + Ksv[Q] \qquad (2.3)$$

where $F_0$ and $F$ are the fluorescence intensities at 640 nm in the absence and presence of cyanide, respectively, $K_{SV}$ is the Stern–Volmer quenching constant, and [Q] is the concentration of analyte quencher. According to Eq. 2.3, the $K_{SV}$ for cyanide was determined to be 1.0 μM. In this case, the lowest detection concentration of cyanide is 200 nM, which is about 14 × lower than the maximum level (2.7 μM) of cyanide in drinking water permitted by the World Health Organization (WHO), strongly suggesting the great potential of Au NCs as a highly sensitive cyanide probe.

**Fig. 2.8** Fluorescence
response of BSA-stabilized
Au NCs upon addition of
different concentrations of
cyanide. The insert is the $F_0/F$
value as a function of the
concentration of cyanide.
(Copyright 2010 Wiley-VCH)

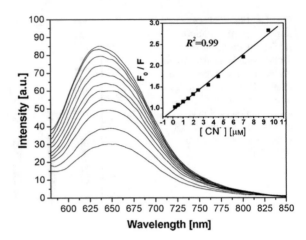

## 2.3.4  Specificity Analysis

To investigate whether our system is specific for cyanide, we measured the
fluorescence response of this sensing system with eighteen common anions under
the same conditions including $S^{2-}$, $SCN^-$, $Br^-$, $CO_3^{2-}$, $Cl^-$, $NO_3^-$, $SO_4^{2-}$, $N_3^-$,
$CN^-$, $EDTA^{2-}$, $Ac^-$, $PO_4^{3-}$, $NO_2^-$, $IO_3^-$, citrate, $C_2O_4^{2-}$, $F^-$, and $I^-$. As shown in
Fig. 2.9, only cyanide could induce a drastic decrease in the fluorescence intensity,
whereas, no obvious fluorescence changes were observed in the presence of any
other anion. The tolerance concentrations of these anions were at least 20 × the
$CN^-$ concentration for detecting $CN^-$ using Au NCs. The above results indicate that
our sensing system is highly selective towards cyanide over other anions. This
excellent selectivity can be attributed to the aforementioned Elsner reaction between
the cyanide and Au atoms. In contrast, for currently studied fluorescence-based
cyanide-sensing systems that are generally based upon the strong nucleophilicity of
cyanide towards the sensor, the strong competition of sensors with cyanide and
fluoride may result in poor selectivity, as fluoride is also a strong nucleophile.
Moreover, the response of Au NCs to various environmentally relevant metal ions
was also investigated (Fig. 2.10). The results showed excellent selectivity for $CN^-$
over $K^+$, $Na^+$, $Mg^{2+}$, $Ca^{2+}$, and $Al^{3+}$. Also, in the presence of a chelating ligand,
2,6-pyridinedicarboxylic acid (PDCA), BSA and glutathione, Au NCs showed high
selectivity against $Zn^{2+}$, $Ni^{2+}$, $Cd^{2+}$, $Hg^{2+}$, $Pb^{2+}$, $Cu^{2+}$, $Co^{2+}$, and $Ag^+$. All these
results revealed the great potential of Au NCs for highly specific detection of
cyanide in the real samples, such as drinking water or biological samples.

**Fig. 2.9 a** Fluorescence response of BSA-stabilized Au NCs to different anions at the concentration of 5 μM at pH 12. **b** $F_0/F$ plotted against 5 μM $CN^-$ with the coexistence of other anions at a concentration of 100 μM and a pH of 12. (Copyright 2010 Wiley-VCH)

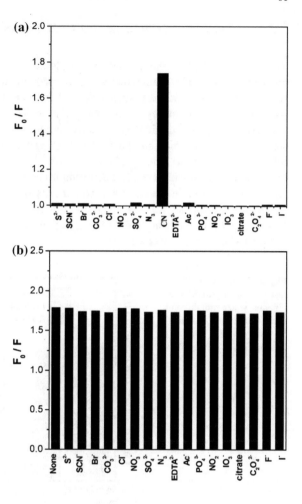

**Fig. 2.10** Fluorescence response of BSA-stabilized Au NCs to different cations in the presence of chelating ligand, PDCA (2 mM), BSA (0.4 mg/mL), and GSH (0.4 mM) ($CN^-$: 5 μM, $Fe^{3+}$: 2 μM, $Zn^{2+}$, $Ni^{2+}$, $Cd^{2+}$, $Hg^{2+}$, $Pb^{2+}$, $Cu^{2+}$, $Co^{2+}$; $Ag^+$: 0.1 μM; X: 0.01 mM; X represents the mixture of $K^+$, $Na^+$, $Li^+$, $Mg^{2+}$, $Ca^{2+}$, $Ba^{2+}$ and $Al^{3+}$). (Copyright 2010 Wiley-VCH)

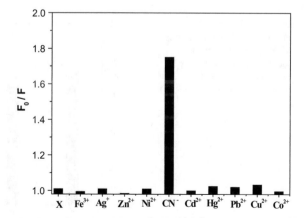

### 2.3.5  Cyanide Detection in the Real Samples

To evaluate whether the Au-NC-based fluorescent sensor developed here is applicable to natural systems, real water samples, including local groundwater, tap water, pond water, and lake water collected from South Lake at Changchun City, were analyzed using our cyanide-sensing system. The experimental results showed that this sensing system gives no obvious fluorescence response to the above water samples, suggesting that, similar to deionized water, these real water samples had little interference in the performance of this sensing system (Fig. 2.11). Nevertheless, the addition of the water samples spiked with 50 μM cyanide led to a significant decrease in the fluorescence intensity of Au NCs. Interestingly, similar fluorescence response was observed for different cyanide-spiked water samples (Fig. 2.12), further indicating the high selectivity of our sensing system over other compositions in real water samples.

**Fig. 2.11** Fluorescence response of the Au NCs to different water samples and water samples spiked with 50 μM cyanide at a pH of 12. (Copyright 2010 Wiley-VCH)

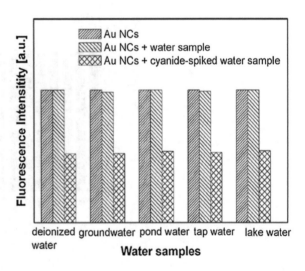

**Fig. 2.12** Fluorescence response of different batches of Au NCs to the water samples spiked with 50 μM cyanide at a pH of 12. (Copyright 2010 Wiley-VCH)

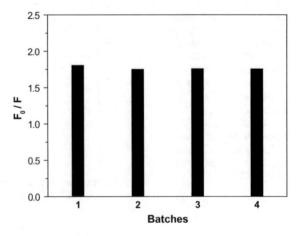

**Table 2.1** Recovery of Au NCs in the detection of cyanide in the real samples. (Copyright 2010 Wiley-VCH)

| $CN^-$ added to the water sample [$\mu$M] | $CN^-$ detected [$\mu$M] | Recovery (%) |
|---|---|---|
| 2 | 1.97 | 98.5 |
| 4 | 3.72 | 93.0 |
| 6 | 5.90 | 98.3 |
| 8 | 7.82 | 97.7 |

Based on the above results, a standard addition method was performed to determine the concentration of cyanide in the cyanide spiked water samples. Taking the cyanide-spiked lake water as an example, a linear correlation was obtained between $F_0/F$ and the concentration of cyanide over the range from 1 $\mu$M to 9 $\mu$M ($R^2 = 0.98$). The recoveries were 93–98.5% (Table 2.1), demonstrating that this novel sensing system has great potential for quantitative analysis of cyanide levels in the environmental samples.

## 2.4 Conclusions

In summary, in this chapter, we have described an Au-NC-based fluorescent sensor for highly sensitive and selective detection of cyanide in aqueous solution, based on the cyanide etching-triggered fluorescence quenching of the Au NCs. Different from traditional detection methods, this method is relatively simple, involving no complex organic synthesis and complicated instruments, while enabling a high sensitivity and excellent selectivity toward cyanide over other common anions. Also, this method can work directly in aqueous solution and does not require toxic organic reagents as an assistant solvent. These unique features make this simple, cost-effective sensing system very attractive for reliable detection of cyanide in real samples such as food, soil, water and biological samples.

## References

1. Alcorta R (2004) Smoke inhalation & acute cyanide poisoning. Hydrogen cyanide poisoning proves increasingly common in smoke-inhalation victims. JEMS 29:6–15
2. Baud FJ, Barriot P, Toffis V et al (1991) Elevated blood cyanide concentrations in victims of smoke inhalation. Engl J Med 325:1761–1766
3. Vennesland B, Comm EE, Knownles CJ et al (1981) Cyanide in biology. Academic, London
4. Sun H, Zhang YY, Si SH et al (2005) Piezoelectric quartz crystal (PQC) with photochemically deposited nano-sized Ag particles for determining cyanide at trace levels in water. Sens Actuators B Chem 108:925–932
5. Day JK, Bresner C, Coombs ND et al (2008) Colorimetric fluoride ion sensing by polyborylated ferrocenes: structural influences on thermodynamics and kinetics. Inorg Chem 47:793–804

6. Palomares E, Martínez-Diaz MV, Torres T et al (2006) A highly sensitive hybrid colorimetric and fluorometric molecular probe for cyanide sensing based on a subphthalocyanine dye. Adv Funct Mater 16:1166–1170
7. Tomasulo M, Sortino S, White AJP et al (2006) Chromogenic oxazines for cyanide detection. Org Chem 71:744–753
8. Anzenbacher P, Tyson DS, Jursíková K et al (2002) Luminescence lifetime-based sensor for cyanide and related anions. J Am Chem Soc 124:6232–6233
9. Chung YM, Raman B, Kim DS et al (2006) Fluorescence modulation in anion sensing by introducing intramolecular H-bonding interactions in host–guest adducts. Chem Commun 2008:186–188
10. Ekmekci Z, Yilmaz MD, Akkaya EU (2008) A monostyryl-boradiazaindacene (BODIPY) derivative as colorimetric and fluorescent probe for cyanide ions. Org Lett 10:461–464
11. Li ZA, Lou XD, Yu HB et al (2008) An imidazole-functionalized polyfluorene derivative as sensitive fluorescent probe for metal ions and cyanide. Macromolecules 41:7433–7439
12. Zeng Q, Cai P, Li Z et al (2008) An imidazole-functionalized polyacetylene: convenient synthesis and selective chemosensor for metal ions and cyanide. Chem Commun 2008: 1094–1096
13. Lee KS, Kim HJ, Kim JH et al (2008) Fluorescent chemodosimeter for selective detection of cyanide in water. Org Lett 10:49–51
14. Chung Y, Lee H, Ahn KH (2006) Ratiometric fluorescence detection of cyanide based on a hybrid coumarin–hemicyanine dye: the large emission shift and the high selectivity. J Org Chem 71:9470–9474
15. Shang L, Dong SJ (2009) Design of fluorescent assays for cyanide and hydrogen peroxide based on the inner filter effect of metal nanoparticles. Anal Chem 81:1465–1470
16. Xie J, Zheng Y, Ying JY (2009) Protein-directed synthesis of highly fluorescent gold nanoclusters. J Am Chem Soc 131:888–889
17. Scott D, Toney M, Muzikár M (2008) Harnessing the mechanism of glutathione reductase for synthesis of active site bound metallic nanoparticles and electrical connection to electrodes. J Am Chem Soc 130:865–874
18. Jungreis E (1969) Microdetermination of cyanides by atomic absorption spectroscopy. Israel J Chem 7:583–584
19. Wang XB, Wang YL, Yang J et al (2009) Evidence of significant covalent bonding in Au (CN)(2)(-). J Am Chem Soc 131:16368–16370
20. Lakowicz JR (1999) Principles of fluorescence spectroscopy. Kluwer Adcademic & Plenum Press, New York

# Chapter 3
# Application of Gadolinium-Doped Zinc Oxide Quantum Dots for Magnetic Resonance and Fluorescence Imaging

**Abstract** Although the fluorescence imaging (FI) technique has high sensitivity, its penetration capability is very limited. In contrast, magnetic resonance imaging (MRI) has deep tissue penetration but low sensitivity. Thus, more available and accurate diagnostic information can be anticipated after combination of FI with MRI. However, currently existent methods for the fabrication of MRI-FI nanoprobes are complex, and the resultant MRI-FI nanoprobes have many disadvantages such as high toxicity, large particle size, or low relaxivity and quantum yield. To address these issues, we described in this chapter a straightforward and versatile method to develop MRI-FI dual modality nanoprobes by doping $Gd^{3+}$ ions in low toxic ZnO quantum dots (QDs). The resultant Gd-doped ZnO QDs are ultrasmall in size and have enhanced fluorescence resulting from the Gd doping. In vitro experiments confirm that Gd-doped ZnO QDs can successfully label the HeLa cells in short time and present no evidence of toxicity or adverse effects on cell growth. Besides, they exert a strong positive contrast effect with a large longitudinal relaxivity much higher than that of Gd-based clinical MRI contrast agent.

## 3.1 Introduction

Fluorescence imaging (FI) has high sensitivity with the capability of analysis on the molecular level, and has received great attention for biological applications such as in vivo imaging. However, the tissues of interest are generally located deep in the human body, and thus the low penetration of fluorescence imaging (generally less than millimeters) significantly limits the applications of fluorescence imaging in the clinical diagnosis, in particular for those tissues located deep in the body [1, 2].

In contrast, MRI is generally recognized as one of the most powerful imaging tools in the clinical diagnosis with deep tissue penetration, minimum invasiveness, and high resolution for soft tissues. It can provide multi-azimuthal and multiple parameter diagnostic information of various diseases in the patients [3, 4]. Despite these promising features, MRI has low detection sensitivity and can only detect the

© Springer Nature Singapore Pte Ltd. 2018
Y. Liu, *Multifunctional Nanoprobes*, Springer Theses,
DOI 10.1007/978-981-10-6168-4_3

target with the size over several micrometers, which makes it hard for early detection of the diseases, such as cancers [5–7].

Given these complementary features between FI and MRI, it is possible to simultaneously realize high sensitivity and deep tissue penetration when combine the FI and MRI, providing more accurate diagnostic information and thus representing a promising research field in the biomedicine. Nanoparticles offer a promising strategy to achieve this goal, as they can incorporate multiple functionalities into a single nanoformulation. In the past decades, varieties of nanoparticle-based FI/MRI dual imaging probes have been investigated for biomedical applications. They have been divided into two categories: T2-weighted MRI/FI nanoprobes and T1-weighted MRI/FI nanoprobes. The first one involves incorporating paramagnetic ions into quantum dots (QDs) such as CdSe or coating a magnetic core and fluorophores within the shells such as silica [8–14]. Despite high imaging performance, T2-weighted imaging, which decreases the signals, generally suffers from confusion with the reduced signals from bleeding, calcification or metal deposits [15, 16]. To overcome this issue, T1-weighted MRI/FI nanoprobes have moved to spotlight. These nanoprobes were usually prepared by doping the QDs with paramagnetic ions or covalently conjugating of paramagnetic materials to the silica-coated QDs [17–19]. Recently, Yong Il Park et al. have developed a novel T1-weighted MRI/FI nanoprobe by coating the upconversion fluorescent NPs with a $NaGdF_4$ shell. Rare earth ions such as $Tb^{3+}$ or $Eu^{3+}$ have also been doped into the Gd-containing NPs for T1-weighted MRI/FI imaging [20, 21].

Despite promising imaging capabilities, these strategies post several limitations. For example, silica coating makes the synthesis more complicated and less reproducible, and usually generates large NPs. From the viewpoint of practical applications, larger particles were not very suitable for biological and medical fields, in particular for labeling functional subcellular or proteins, as larger particles often have an influence on their biological function and are more likely to be recognized and cleared by the phagocyte. On the other hand, the upconversion fluorescence NPs have low quantum yields (OYs) and low relaxivity after doping of Gd. Although CdSe QDs have high photostability and quantum yields, the potential toxicity associated with the release of $Cd^{2+}$ and $Se^{2+}$ ions represents another intractable problem for in vivo applications [22, 23]. Thus, it is highly desirable to develop new MRI/FI nanoprobes with a high quantum yield, high relaxivity, low toxicity, good photostability, and facile preparation processes.

To overcome these obstacles, in this chapter, we have described a facile method to synthesize a novel T1-weighted MRI/FI nanoprobe based on Gd-doped ZnO QDs, and systemically investigated their optical and magnetic properties, and the potential in the MRI and FI dual imaging.

## 3.2 Experimental Section

### 3.2.1 Materials

Zinc acetate dihydrate was otained from Beijing Chemical Reagent factory (Beijing, China). Tetramethylammonium hydroxide (TMAH) and Gadolinium acetate hydrate were purchased from Alfa Aesar. Oleic acid was obtained from Sigma Aldrich. Other chemicals were of analytical grade and used without any further purification. Water used throughout all experiments was purified with the Millipore system.

### 3.2.2 Synthesis of ZnO QDs and Gd-Doped ZnO QDs

ZnO QDs were synthesized according to a previously reported method [24]. Briefly, $Zn(OAc)_2$ $2H_2O$ (1.2 mM) was first dissolved in 20 mL of ethanol under vigorous stirring. 70 mL of oleic acid was then added and the mixture was heated to reflux. TMAH (0.526 mL) in 5 mL of refluxing ethanol was rapidly injected into this mixture under vigorous stirring. Three minutes later, the reaction was stopped by adding 50 mL of ethanol, followed by cooling down immediately in an ice bath. The oleate-stabilized ZnO QDs were isolated by congregating, washed with ethanol several times, and finally redispersed in 10 mL toluene. Gd-doped ZnO QDs with different x values were synthesized by adding stoichiometric amount of gadolinium acetate to the zinc acetate precursor solution, with constant total initial concentration of metal ions, as shown in Table 3.1.

### 3.2.3 Surface Modification of Gd-Doped ZnO QDs

Under stirring, 1 mL of AEAPS in toluene (0.1 M) was added to the solution of ZnO QDs dispersed in 10 mL toluene at room temperature. After 5 min, 1 mL of TMAH in ethanol (0.2 M) was then injected. The mixture was refluxed at 85 °C for 45 min, couple with cooling down to room temperature. The functional ZnO QDs

**Table 3.1** Detailed concentrations of starting materials for the synthesis of Gd-doped ZnO QDs. (Copyright 2011 Elsevier)

| $Gd(CH_3COO)_3 \cdot xH_2O$ (g) | $Zn(OAc)_2 \cdot 2H_2O$ (g) | Gd/Zn ratio(x) |
|---|---|---|
| 0.0233 | 0.2502 | 0.02 |
| 0.0406 | 0.2371 | 0.04 |
| 0.0699 | 0.2239 | 0.05 |
| 0.09321 | 0.2107 | 0.08 |
| 0.1165 | 0.1976 | 0.10 |
| 0.1844 | 0.1398 | 0.12 |
| 0.233 | 0.1317 | 0.30 |

with amine groups on their surface were obtained by centrifugation and washed several times with toluene, acetone and water. Finally, QDs were redispersed in water and stored at 4 °C in the dark for further characteristics.

### 3.2.4 Cell Cytotoxicity Studies

HeLa cells were incubated in the culture medium at 37 °C in an atmosphere of 5% $CO_2$ and 95% air for 24 h. After removal of the culture medium, fresh medium containing different concentrations of Gd-doped ZnO QDs were added, and the cells were cultured for another 24 h. The cell viabilities were tested by MTT assay.

### 3.2.5 Confocal Imaging

HeLa cells were incubated with Gd-doped ZnO QDs (x = 0.08) at 37 °C in an atmosphere of 5% $CO_2$ and 95% air for different time. Then, the cells were isolated by centrifugation washed with PBS several times, and subsequently redispersed in PBS. Images were collected using confocal microscopy with the excitation at 405 nm.

### 3.2.6 T1-Weighted MRI

For in vitro MRI, Gd-doped ZnO QDs (x = 0.08) were dispersed in water with $Gd^{3+}$ concentrations in the range from 0 to 0.05 mM. HeLa cells were incubated with Gd-doped ZnO QDs (x = 0.08) for 2 h. After centrifugation, the cell pellet was covered with agarose. T1-weighted images of Gd-doped ZnO QDs (x = 0.08) with varied $Gd^{3+}$ concentrations and cells treated and untreated with Gd-doped ZnO QDs were collected using 1.5 T human clinical scanner.

### 3.2.7 Characteristic

The nanoparticle size was examined using JEOL 2000-FX transmission electron microscopy (TEM). The molar ratios of Gd/Zn in the products were determined by an ELAN 9000/DRC ICP-MS system. UV/Vis spectra were recorded on a VARIAN CARY 50 UV/Vis spectrophotometer. The fluorescence spectra were obtained using a PerkineElmer LS 55 luminescence spectrometer. XPS was performed using a VG ESCALAB MKII spectrometer, utilizing the XPSPEAK software (Version 4.1) to deconvolute the narrow-scan XPS spectra of Gd3d of Gd-doped ZnO QDs and $Gd_2O_3$, using adventitious carbon to calibrate the binding energy of C1s (284.5 eV).

## 3.3 Results and Discussion

### 3.3.1 Synthesis and Characteristic of Gd-Doped ZnO QDs

It has been confirmed that zinc acetate could hydrolyze and generated ZnO QDs in the ethanol containing TMAH [24]. The as-prepared ZnO QDs have a high OY and good photostability. In this chapter, we have successfully prepared Gd-doped ZnO QDs by simultaneously introducing gadolinium acetate during the hydrolysis of zinc acetate. Compared to previously reported methods, our strategy is very simple and fast without the need of any coating processes. Fig. 3.1 shows the TEM images of Gd-doped ZnO QDs with different Gd/Zn ratios. All the QDs are spherical in shape and uniform in size. With the increasing Gd ratio, the size of Gd-doped ZnO QDs was decreased. The diameter of the pure ZnO QDs is 6 nm, and the diameter of Gd-doped ZnO QDs decreased to 4 nm when the Gd/Zn ratio is 0.08. Further increase of the Gd/Zn ratio led to smaller QDs, which were too small to be seen with TEM. Meanwhile, selected area electron diffraction analysis presented similar patterns for ZnO QDs and Gd-doped ZnO QDs (x = 0.08 and 0.3), and all the diffraction rings were indexed to wurtzite ZnO. High-resolution TEM images revealed lattice fringes

**Fig. 3.1** TEM images of Gd-doped ZnO QDs with x = 0 (**a**), 0.08 (**b**) and 0.3 (**c**), the insets show the corresponding electron diffraction patterns. (**d**–**f**) The corresponding high-resolution TEM images for **a**–**c**. (Copyright 2011 Elsevier)

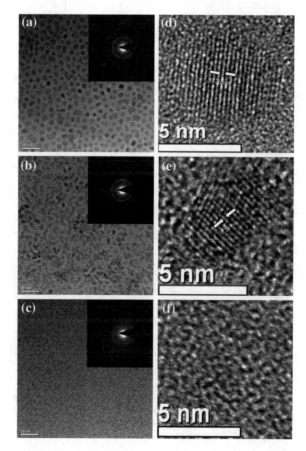

with a spacing of about 0.26 nm for pure ZnO QDs and Gd-doped ZnO QDs ($x = 0.08$), which were well consistent with the lattice spacing in the (002) planes of ZnO wurtzite phase. These results suggested that Gd doping did not induce any significant lattice distortions of ZnO QDs. However, when x increased to 0.3, the crystallinity of the ZnO QDs was very weak and no clear lattice fringe was observed.

To further confirm the crystal structure of the product, X-ray diffraction (XRD) analysis was also performed and the results were shown in Fig. 3.2. All the reflection peaks of undoped ZnO QDs were indexed to the ZnO structure (JCPDS No. 65–3411 and JCPPDS No. 21–1486), while the XRD spectra of Gd-doped ZnO QDs showed the patterns of a single hexagonal phase of ZnO (JCPDS No. 65–3411) with weakened crystallinity as x value increased. Surprisingly, no reflection peak of $Gd_2O_3$ was observed in these XRD patterns, which we hypothesized is due to the formation of amorphous species. In addition, the average particle's diameter calculated according to the DebyeeScherer formula decreased as the increase of x value (Table 3.2), which were in accordance with those obtained from TEM images.

Next, energy-dispersive X-ray analysis (EDAX) and X-ray photoelectron spectroscopy (XPS) analysis were carried out to determine the oxidative states of

**Fig. 3.2** XRD spectra of Gd-doped ZnO QDs with different Gd/Zn ratios. (Copyright 2011 Elsevier)

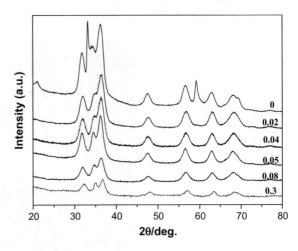

**Table 3.2** Comparison between Gd-doped ZnO QDs with different x values. (Copyright 2011 Elsevier)

| ZnO QDs with different *x* values (ICP data) | Size[a] (nm) | Size[b] (nm) | Band gap[c] (eV) |
|---|---|---|---|
| 0 | 6.0 | 5.7 | 3.42 |
| 0.02 | 5.18 | 5.11 | 3.435 |
| 0.04 | 5.02 | 4.96 | 3.44 |
| 0.05 | 4.39 | 4.36 | 3.47 |
| 0.08 | 4.14 | 3.9 | 3.49 |
| 0.10 | 3.95 | 3.8 | 3.50 |
| 0.12 | 3.76 | 3.56 | 3.52 |
| 0.30 | 3.55 | 3.27 | 3.55 |

[a]Measured by XRD. The average particle size was calculated according to the Debyee Scherer formul)
[b]Determined using Meulenkamp's empircal formula.
[c]measured by UV-Vis spectra

Gd in the Gd-doped ZnO QDs. For undoped ZnO QDs, EDAX pattern showed strong characteristic peaks of Zn, O, and C in the ZnO QDs and no Gd or other impurity elements were observed, whereas remarkable Gd peaks were seen in the EDAX patterns of Gd-doped ZnO QDs, indicating the successful doping of Gd in

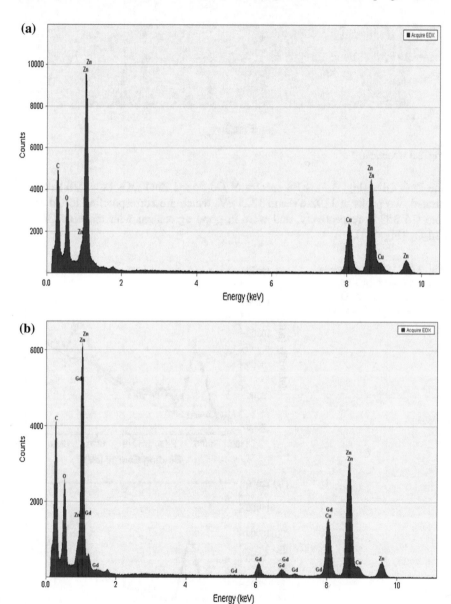

**Fig. 3.3** EDAX spectra of **a** ZnO QDs, **b** Gd-doped ZnO QDs (x = 0.08), and **c** Gd-doped ZnO QDs (x = 0.3). (Copyright 2011 Elsevier)

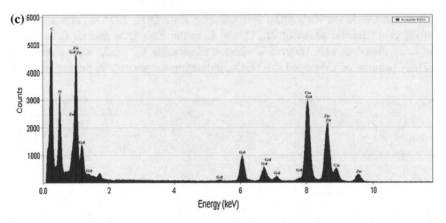

**Fig. 3.3** (continued)

the ZnO QDs (Fig. 3.3). XPS spectra of Gd-doped ZnO QDs (x = 0.08) demonstrated two peaks at 1192 eV and 1225 eV, which are corresponding to Gd 3d5/2 and Gd 3d3/2, respectively, and were in good agreement with those of $Gd^{3+}$ in $Gd_2O_3$ (Fig. 3.4).

**Fig. 3.4** Gd3d XPS spectra of **a** Gd-doped ZnO QDs (x = 0.08) and **b** $Gd_2O_3$. (Copyright 2011 Elsevier)

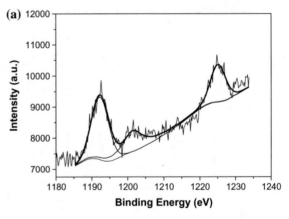

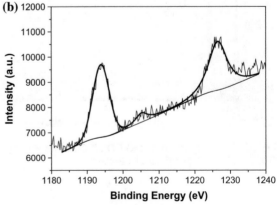

### 3.3.2 Optical Properties of Gd-Doped ZnO QDs

Under irradiation with the 340 nm light, ZnO QDs exhibited a strong yellow emission centered at about 565 nm (Fig. 3.5). Interestingly, we found a significant enhancement in fluorescence intensity and a blue shift of emission band with the increase of the Gd ration in ZnO QDs, achieved to the maximum when the x value is 0.08. The QY value using Rhodamine 6G as reference fluorescent dye enhanced from 13.5% (x = 0) to 34% (x = 0.08), and the maximum emission shifted from 565 nm to 550 nm. However, further increase in the concentration of Gd led to an obvious decrease in fluorescence intensity and a further blue shift.

It has been widely recognized that the emission of ZnO QDs derived from the oxygen vacancies, and the emission intensity increased as a function of the amount of oxygen vacancies [25–27]. It has also been suggested that the decrease of particle

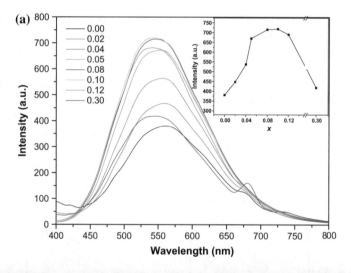

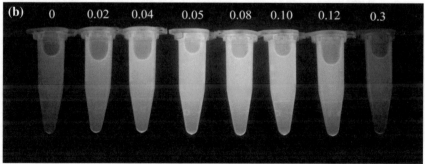

**Fig. 3.5** **a** Fluorescence emission spectra of Gd-doped ZnO QDs with different x values at an excitation wavelength of 340 nm. **b** Fluorescence of Gd-doped ZnO QDs with different x values under UV light at 365 nm. (Copyright 2011 Elsevier)

size could induce the increase of the amount of oxygen vacancies [28]. Thus, we hypothesized that the QY enhancement of ZnO QDs might be resulted from Gd doping induced size decrease of ZnO QDs and subsequent increase of the amount of oxygen vacancies. It should be noted that excessive Gd would consumed the ZnO nanoparticles, thus decreasing the fluorescence intensity.

In general, the blue-shift of fluorescence of ZnO QDs was associated with their band-gap enlargement as a result of the quantum size effect [29, 30]. To examine whether the Gd doping induced the band gap enlargement of ZnO QDs, UV/Vis absorption spectra of the Gd-doped ZnO QDs with different x values were recorded (Fig. 3.6), and their band gaps were calculated according to the following equation, given the fact that ZnO is a direct band gap semiconductor [31]:

$$\alpha = A \bullet \frac{(h\nu - E)^{1/2}}{h\nu} \tag{3.1}$$

where h, E and A are absorption coefficient, band gap and constant, respectively. According to Eq. 3.1, the optical band gaps (Eg) of ZnO QDs with different x values can be estimated by extrapolating the linear part near the onset in a plot of $(ah\nu)^2$ versus $h\nu$. A clear blue shift in band gap was noticed as the x value increased (Fig. 3.6 insert). As seen in Fig. 3.6, the band gap of ZnO QDs increased with the increasing Gd ratio. For example, the band gaps of ZnO QDs with x = 0.08 and 0.3 increased to 3.49 eV and 3.55 eV, respectively, compared to the undoped ZnO QDs (3.42 eV). Consequently, it was reasonable to conclude that the blue shift in the emission of ZnO QDs was due to the band-gap widening effect triggered by the Gd

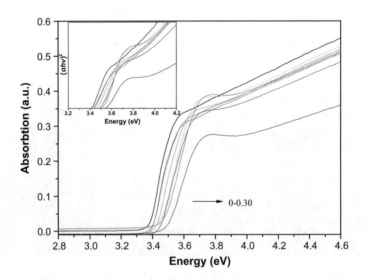

**Fig. 3.6** UV-Vis spectra of Gd-doped ZnO QDs with different x values. (Copyright 2011 Elsevier)

doping. In turn, the particle's size was further determined as 5.7 nm (x = 0) and 3.9 nm (x = 0.08) according to the Meulenkamp's empirical equation (Eq. 3.2):

$$1240 = \lambda_{1/2}[a + \frac{b}{D^2} - \frac{c}{D}] \qquad (3.2)$$

Where $\lambda_{1/2}$ is the value of half absorbance, a, b, and c are constants, and D is the diameter of the particle. The results are shown in Table 3.2, which was well consistent with the results from TEM images and XRD analysis.

### 3.3.3 Cytotoxicity Studies

Given the significantly enhanced fluorescence of Gd-doped ZnO QDs (x = 0.08), we proceeded to evaluate their potential application as imaging probes in vivo. The cytotoxicity of Gd-doped ZnO QDs was first investigated using the MTT assay (Fig. 3.7). Prior to the test, Gd-doped ZnO QDs were modified with N-(2-aminoethyl) aminopropyltrimethoxysilane (AEAPS), considering that the amine groups on the surface of QDs favored their dispersion in aqueous solutions and simultaneously provided a functionalized surface. As expected, the cell viability was not hindered when cells were treated with the Gd-doped ZnO QDs at the concentration up to 1 mM. Moreover, the cell viability remained ~ 50% at the highest dose of 5 mM. In contrast, some traditional Cd-based QDs showed severe toxicity even at a very low concentration [32–34]. For instance, a cell viability loss of 50% was observed when the procine renal proximal cells incubated with PEGylated CdSe/ZnS QDs [35]. All these results demonstrated the low toxicity of our Gd-doped ZnO QDs. Another concern lies in the possible release of free $Gd^{3+}$ ions from Gd-doped QDs due to the high toxicity of free $Gd^{3+}$ ions. With ICP analysis, we did not test obvious $Gd^{3+}$ leaching from the Gd-doped ZnO QDs after

**Fig. 3.7** Cytotoxicity of Gd-doped ZnO QDs (x = 0.08). (Copyright 2011 Elsevier)

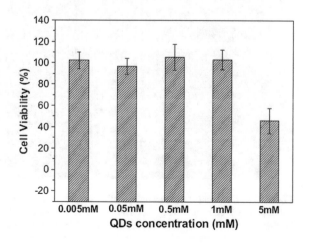

dialysis against PBS buffer for one week, indicating their stability under the physiological environment.

### 3.3.4   In Vitro Fluorescence Imaging

To assess the potential of Gd-doped ZnO QDs in the fluorescence imaging, HeLa cells were incubated with Gd-doped ZnO QDs for different time intervals, and the fluorescence was examined by confocal imaging. As shown in Fig. 3.8, after incubation with Gd-doped ZnO QDs for 30 min, a bright yellow emission was observed in the cells, indicating Gd-doped ZnO QDs could be used as effective labeling probes. In addition, the small size of Gd-doped ZnO QDs could efficiently decrease the cytophagy of phagocyte, potentially providing a sufficiently long time for labelling.

### 3.3.5   In Vitro MR Imaging

Given the fact that $Gd^{3+}$ ions can accelerate longitudinal (T1) relaxation of water protons and exert bright contrast in regions where the nanoprobes localize, we next

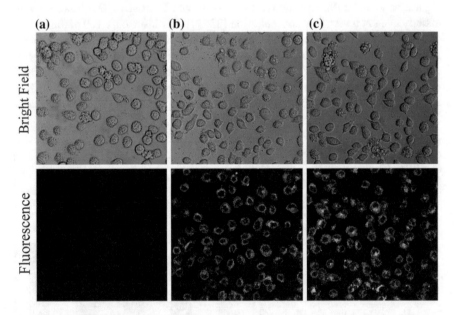

**Fig. 3.8 a** Confocal laser scanning microscopic images of HeLa cells incubated without Gd-doped ZnO QDs (x = 0.08). **b, c** HeLa cells incubated with Gd-doped ZnO QDs (x = 0.08) for 30 min and 2 h at the same concentration (0.625 mM). (Copyright 2011 Elsevier)

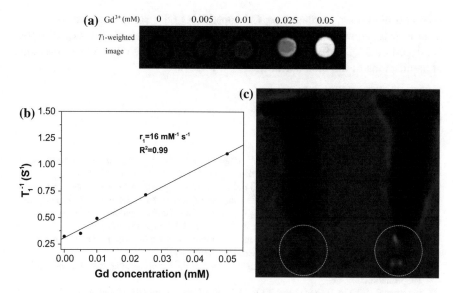

**Fig. 3.9** **a** T1-weighted magnetic resonance image of Gd-doped ZnO QDs (x = 0.08) at different concentrations scanned with a 1.5 T clinical MRI system. **b** The linear relationship between T1 relaxation rates (1/T1) and Gd concentrations for Gd-doped ZnO QDs (x = 0.08). **c** T1-weighted image of blank HeLa cells pellet (left) and HeLa cells incubated with Gd-doped ZnO QDs at 0.01 M Gd for 2 h. (Copyright 2011 Elsevier)

investigated the MRI capability of Gd-doped ZnO QDs using a 1.5 T human clinical scanner. Fig. 3.9a is the T1-weighted MRI image of Gd-doped ZnO QDs at different concentrations of Gd. Gd-doped ZnO QDs showed a linear increase in the MRI signal with increasing Gd concentrations. The specific relaxivity value (r1) was determined to be 16 mM$^{-1}$ S$^{-1}$, which is much larger than that of the commercial Magnevist (< 5 mM$^{-1}$ S$^{-1}$). Next, T1-weighted MR imaging was also performed on the cell pellets under the same magnetic strength. The T1-weighted MR image of HeLa cells treated with Gd-doped ZnO QDs was brighter than that of untreated HeLa cells. All these results clearly suggested that the Gd-doped ZnO QDs could serve as effective T1-MRI contrast agents.

## 3.4   Conclusion

In this chapter, we have developed a facile strategy for the fabrication of a novel MRI-FI dual imaging nanoprobe based on ultrasmall Gd-doped ZnO QDs. These Gd-doped ZnO QDs exhibited significantly enhanced yellow emission after Gd doping. The use of Gd-doped ZnO QDs as in vitro optical imaging agents has been demonstrated in HeLa cells, and the result showed that Gd-doped ZnO QDs succefully labeled the HeLa cells in short time without inducing obvious cytotoxicity even at a high dose. Furthermore, Gd-doped ZnO QDs showed contrast enhancement in MRI compared to the commercial MRI contrast agent. We expected that Gd-doped ZnO QDs after modification with target ligands would find a broad range of applications in the biomedical field.

# References

1. Tsai CP, Hung Y, Chou YH et al (2008) High-contrast paramagnetic fluorescent mesoporous silica nanorods as a multifunctional cell-imaging probe. Small 4:186–191
2. Setua S, Menon D, Asok A et al (2010) Folate receptor targeted, rare-earth oxide nanocrystals for bi-modal fluorescence and magnetic imaging of cancer cells. Biomaterials 31:714–729
3. Mulder WJM, Strijkers GJ, Tilborg GAF van et al (2006) Lipidbased nanoparticles for contrast-enhanced MRI and molecular imaging. NMR Biomed 19:142–164
4. Lauffer RB et al (1987) Paramagnetic metal complexes as water proton relaxation agents for NMR imaging: theory and design. Chem Rev 87:901–927
5. Reid T (2009) Magnetic resonance imaging: Forcing the nanoscale. Nat Nanotechnol. doi:10.1038/nnano.2009.14
6. Caravan P, Ellison JJ, McMurry TJ et al (1999) Gadolinium (III) chelates as MRI contrast agents: structure, dynamics, and applications. Chem Rev 99:2293–2352
7. Anderson EA, Isaacman S, Peabody DS et al (2006) Viral nanoparticles donning a paramagnetic coat: Conjugation of MRI contrast agents to the MS2 capsid. Nano Lett 6:1160–1164
8. Kim J, Kim HS, Lee N et al (2008) Multifunctional uniform nanoparticles composed of a magnetite nanocrystal core and a mesoporous silica shell for magnetic resonance and fluorescence imaging and for drug delivery. Angew Chem Int Ed 47:8438–8441
9. Yi DK, Selvan ST, Lee SS et al (2005) Silica-coated nanocomposites of magnetic nanoparticles and quantum dots. J Am Chem Soc 127:4990–4991
10. Hu KW, Hsu KC, Yeh CS (2010) pH-Dependent biodegradable silica nanotubes derived from Gd(OH)$_3$ nanorods and their potential for oral drug delivery and MR imaging. Biomaterials 31:6843–6848
11. Schooneveld MM van, Vucic E, Koole R et al (2008) Improved biocompatibility and pharmacokinetics of silica nanoparticles by means of a lipid coating: A multimodality investigation. Nano Lett 8:2517–2525
12. Selvan ST, Patra PK, Ang CY et al (2007) Synthesis of silica-coated semiconductor and magnetic quantum dots and their use in the imaging of live cells. Angew Chem Int Ed 46:2448–2452
13. Lee JE, Lee N, Kim H et al (2010) Uniform mesoporous dyedoped silica nanoparticles decorated with multiple magnetite nanocrystals for simultaneous enhanced magnetic resonance imaging, fluorescence imaging, and drug delivery. J Am Chem Soc 132:552–557
14. Santra S, Bagwe RP, Dutta D et al (2005) Synthesis and characterization of fluorescent, radio-opaque, and paramagnetic silica nanoparticles for multimodal bioimaging applications. Adv Mater 17:2165–2169
15. Werner EJ, Datta A, Jocher CJ et al (2008) High-relaxivity MRI contrast agents: Where coordination chemistry meets medical imaging. Angew Chem Int Ed 47:8568–8580
16. Bulte JWM, Kraitch-man DL (2004) Iron oxide MR contrast agents for molecular and cellular imaging. NMR Biomed 17:484–499
17. Santra S, Yang H, Holloway PH et al (2005) Synthesis of water-dispersible fluorescent, radio-opaque, and paramagnetic CdS:Mn/ZnS quantum dots: A multifunctional probe for bioimaging. J Am Chem Soc 127:1656–1657
18. Wang S, Jarrett BR, Kauzlarich SM et al (2007) Core/shell quantum dots with high relaxivity and photoluminescence for multimodality imaging. J Am Chem Soc 129:3848–3856

19. Li IF, Yeh CS (2010) Synthesis of Gd doped CdSe nanoparticles for potential optical and MR imaging applications. J Mater Chem 20:2079–2081
20. Ju Q, Tu D, Liu Y et al (2012) Amine-functionalized lanthanide-doped $KGdF_4$ nanocrystals as potential optical/magnetic multimodal bioprobes. J Am Chem Soc 134:1323–1330
21. Petoral RM Jr, Soderlind F, Klasson A et al (2009) Synthesis and characterization of $Tb^{3+}$-doped $Gd_2O_3$ nanocrystals: A bifunctional material with combined fluorescent labeling and MRI contrast agent properties. J Phys Chem C 113:6913–6920
22. Kirchner C, Liedl T, Kudera S (2005) Cytotoxicity of colloidal CdSe and CdSe/ZnS nanoparticles. Nano Lett 5:331–338
23. Hardman R (2006) Toxicologic review of quantum dots: Toxicity depends on physicochemical and environmental factors. Environ Health Perspect 114:165–172
24. Jana NR, Yu HH, Ali EM et al (2007) Controlled photostability of luminescent nanocrystalline ZnO solution for selective detection of aldehydes. Chem Commun 14:1406–1408
25. Vanheusden K, Warren WL, Seager CH et al (1996) Mechanisms behind green photoluminescence in ZnO phosphor powders. J Appl Phys 79:7983–7990
26. Dijken AV, Meulenkamp EA, Vanmaekelbergh D et al (2000) Identification of the transition responsible for the visible emission in ZnO using quantum size effects. J Lumin 90:123–128
27. Dijken AV, Meulenkamp EA, Vanmaekelbergh D et al (2000) The kinetics of the radiative and nonradiative processes in nanocrystalline ZnO particles upon photoexcitation. J Phys Chem B 104:1715–1723
28. Xiong HM, Shchukin DG, Mohwald H et al (2009) Sonochemical synthesis of highly luminescent zinc oxide nanoparticles doped with magnesium(II). Angew Chem Int Ed 48:2727–2731
29. Spanhel L, Anderson MA (1991) Semiconductor clusters in the sol-gel process: quantized aggregation, gelation, and crystal growth in concentrated zinc oxide colloids. J Am Chem Soc 113:2826–2833
30. Zhang L, Yin L, Wang C et al (2010) Origin of visible photoluminescence of ZnO quantum dots: Defect-dependent and size-dependent. J Phys Chem C 114:9651–9658
31. Monticone S, Tufeu R, Kanaev AV (1998) Complex nature of the UV and visible fluorescence of colloidal ZnO nanoparticles. J Phys Chem B 102:2854–2862
32. Zhang LW, Yu WW, Colvin VL et al (2008) Biological interactions of quantum dot nanoparticles in skin and in human epidermal keratinocytes. Toxicol Appl Pharmacol 228:200–211
33. Bhang SH, Won N, Lee TJ et al (2009) Hyaluronic acid quantumdot conjugates for in vivo lymphatic vessel imaging. ACS Nano 3:1389–3198
34. Wu C, Shi L, Li Q et al (2010) Probing the dynamic effect of Cys-CdTe quantum dots toward cancer cells in vitro. Chem Res Toxicol 23:82–88
35. Stern ST, Zolnik BS, McLeland CB et al (2008) Induction of autophagy in porcine kidney cells by quantum dots: A common cellular response to nanomaterials? Toxicol Sci 106:140–152

# Chapter 4
# High-Perform Yb-Based Nanoparticulate X-Ray CT Contrast Agent

**Abstract** X-ray computed tomography (CT) has been considered to be the most powerful diagnostic tool in clinical diagnosis due to its many advantages compared to other molecular imaging. X-ray CT contrast agents currently used in clinic scanning are mainly based on iodinated small molecules. However, these small molecules suffer from many disadvantages, such as low contrast efficiency, very short circulation lifetime, and potential renal toxicity. Moreover, some patients are hypersensitive to iodine. These disadvantages have significantly restricted the applications of X-ray CT in biomedicine, particularly in targeted imaging. In this chapter, we describe a first-in-class Yb-based nanoparticulate CT contrast agent. Owing to the attenuation characteristics of Yb, which is matched with the X-ray photon energy used in clinical applications, the Yb-based nanoparticulate CT contrast agent offers a much higher contrast efficacy compared to the clinical iodinated agent at 120 kVp. Along with long circulation time and low toxicity in vivo, these nanoparticles can act as a high-performance CT contrast agent for in vivo angiography and bimodal imageguided lymph node mapping. By doping Gd into the nanoparticles, this contrast agent also shows enhanced upconversion luminescence and MRI capability.

## 4.1 Introduction

Despite high resolution for soft tissues and non-invasiveness, MRI has also some limitations. For example, MRI is not suitable for those patients with a cardiac pacemaker due to the presence of metals. Moreover, MRI scanning is time consuming, and has low apstial resolution. In addition, MRI can not be used for bone detection. X-ray CT imaging represents another powerful imaging tool and widely used for clinic diagnosis. Compared to MRI, X-ray CT imaging is time-saving, and has deep tissue penetration and high resolution. It can provide multiple 3D tissue information after construction by computer [1–3].

Owing to the poor contrast between different soft tissues, contrast agent is generally injected into patient to improve the contrast. Currently, small iodinated

© Springer Nature Singapore Pte Ltd. 2018
Y. Liu, *Multifunctional Nanoprobes*, Springer Theses,
DOI 10.1007/978-981-10-6168-4_4

molecules are routinely used for in vivo contrast enhancement in the clinical setting with an emphasis on their cost-effectiveness rather than performance. In order to provide adequate contrast, it is necessary to inject large doses of these agents, which may lead to adverse effects in patients due to their short circulation lifetime (several seconds to minutes). Moreover, the short circulating time and complexicty in modification make them hard for targeting imaging or other applications that require long blood circulation. In addition, some patients are allergic to iodine and thus not be suitable for CT imaging. Thus, it is highly desirable to development novel X-ray CT contrast agents to overcome these issues.

In addition to the long blood circulation, high contrast efficiency is required for contrast agents to decrease the injection dose and consequently reduce the potential side effects. In our previous work, we have developed a bismuth sulfide-based nanoparticluate CT contrast agent with prolonged circulation time in the blood and good in vivo imaging performance [4]. However, we noticed that similar to other heavy metal-based CT contrast agents, bismuth sulfide did not showed significantly enhanced contrast efficiency compared to the commercial iodine-based small molecules at the same molar concentrations.

When taking a look at the CT imaging mechanism, we will find that these heavy metal elements and iodine are not the optimal contrast element. Thus, we turned our attention to CT imaging mechanism aiming to develop a new generation of CT contrast agent with significantly enhanced contrast efficiency. When the substance is exposed to X-ray, the absorption and reflection of X-ray by the substance will take place, a process referred to as X-ray attenuation. X-ray CT imaging is based on such attenuation mechanism and is dependent on the attenuation coefficient. In human body, different tissues have different attenuation coefficients. A high contrast will be observed if there is obvious attenuation difference among tissues. For example, bone has a much higher attenuation coefficient compared to soft tissues, thus showing intrinsic high contrast without the need of contrast agents. However, the attenuation difference among different soft tissues is too small to distinguish, thereby requiring the contrast agents to improve the contrast among them for diagnosis [5–7].

In general, X-ray attenuation is a result of three interactions occurring between X-ray photons and the traversed matter: coherent scattering, photoelectron effect, and Compton scattering. Coherent scattering is generally so small that its contribution to X-ray attenuation is usually neglected. Compton scattering usually increase the noise and decrease the contrast. In practical applications, low-energy X-ray radiation leads to exposure of the patients to a high dose of radiation energy, as most of the X-rays will be absorbed. A proper increase in the X-ray energy can reduce the exposure dose. In this case, the contribution of Compton scattering to X-ray attenuation will diminish. Hence, the photoelectron effect is the predominant contributor, and its contribution will be higher with absorbers which have higher atomic numbers. Also, the photoelectron effect yields a sharp increase in the mass attenuation coefficient at the K-shell electron binding energy (K-edge), and each element has its own characteristic K-edge value. Thus, the K-edge energy of a better contrast element must match with the X-ray spectrum to guarantee high X-ray attenuation. It is known that an X-ray spectrum generated at 150 KVp shows a maximum intensity around 45–70 keV, and the characteristic radiation emitted is between 57 and 69 keV. Given these information, the X-ray attenuation of contrast agents will be dramatically enhanced by selecting the elements with higher atomic numbers and K-edge values within the highest energy intensity region of the X-ray spectrum.

According to this mechanism, Yb-based nanostructures hold a great promise as CT contrast agents compared to currently investigated Au-, Pt-, Ta-, and Bi-based CT contrast agents. Yb-based CT contrast agents have several unique features including: i) the K-edge energy of Yb (61 keV) well matches with the X-ray spectrum used in clinical CT, thus enabling both higher intrinsic contrast and lower radiation exposure to the patients [8, 9]; ii) low toxicity when encapsulated in the stable nanoparticles [10]; iii) the highest abundance in the earth's crust among the aforementioned metals with the potential for industrial production; and iv) Yb is a well-known component of upconversion luminescence nanocrystals, thus providing a particularly useful platform for the design of multimodal imaging nanoprobes without involving additional modification with other functionalities. Nevertheless, to the best of our knowledge, Yb-based nanoparticulate CT contrast agents remain elusive.

In this chapter, we have successfully developed Yb-containing upconversion nanoparticles (UCNPs) as a new generation of CT contrast agents and systemically evaluated their CT contrast efficiency and in vivo imaging performance.

## 4.2 Experimental Section

### 4.2.1 Materials

$Yb_2O_3$, $Er_2O_3$, $Tm_2O_3$ and $Gd_2O_3$ (the purity > 99.99%) were purchased from Changchun Hepalink rare earth Co. Ltd.. Oleic acid and octadecene were purchased from Sigma-Aldrich. Iobitridol was purchased from Guerbet (France). 1,2- distearoyl-sn-glycero-3-phosphoethanolamine-N-{methoxy-[poly(ethylene glycol)]-2000} (DSPE-PEG2000) was purchased from shanghai advanced viecle technology l.t.d. co (China). Other chemicals were analytical grade and used as received without further purification.

### 4.2.2 Synthesis of OA-Stabilized UCNPs

$Yb_2O_3$, $Er_2O_3$, and $Gd_2O_3$ were separately dissolved in excess 1:1 hydrochloric acid aqueous solution at 80 °C. After evaporation of the hydrochloric acid and water at 140 °C, the resulting powders were redispersed in water to yield the $YbCl_3$ (1M),

ErCl$_3$ (0.1M), and GdCl$_3$ (1M) aqueous stock solutions. For the synthesis of NaYbF$_4$:2%Er nanoparticles, 1.47 mL YbCl$_3$ aqueous solution and 0.3 mL ErCl$_3$ aqueous solution were added in a 50 mL flask. After removal of the water, 15 mL oleic acid and 23 mL octadecene (ODE) were added, followed by heating to 160 °C under Ar protection to yield a homogeneous solution. After cooling down to room temperature, 15 mL methanol solution containing NH$_4$F (0.222 g, 6 mmol) and NaOH (0.15 g, 3.75 mmol) was slowly added. Under vigorous stirring, the solution was slowly heated to 220 °C to remove the residual water and impurities with low boiling point under Ar protection within 30 min. Thereafter, the temperature was increased to 300 °C at a rate of 10 °C/min and remained at this temperature for 1.5 h. After cooling to room temperature in ambient conditions, the nanoparticles were isolated through centrifugation at 10,000 rpm for 20 min and washed with ethanol twice. Finally, the nanoparticles were redispersed in chloroform for further surface modification. The Gd-doped NaYbF$_4$:Er nanoparticles were synthesized with the same procedure by replacing stoichiometrically of YbCl$_3$ with GdCl$_3$ aqueous stock solutions.

### 4.2.3　Surface Modification of OA-Stabilized UCNPs

Water-soluble UCNPs were prepared according to the reported method. In brief, 10 mL of UCNP dispersion in chloroform (10 mg/mL) was added into 20 mL of DSPE-PEG2000 dispersion in chloroform (5 mg/mL). After stirring for 10 min, the chloroform was slowly evaporated and the mixture was maintained at 60 °C for one hour under vacuum. 10 mL of water was then added. After filtration using a 0.2 mm cellulose acetate syringe filter, the excess PEG was removed by centrifugation and washed with water several times.

### 4.2.4　Leaching Study of Yb$^{3+}$ and Gd$^{3+}$ Ions

10 ml of PEG-UCNPs solution (20 mg/mL) was respectively added into dialysis bags (10 kDa cut-off) and dialyzed against blood serum and physiological saline for one weak under general stirring. 0.5 mL of blood serum or physiological saline was taken out, and the concentrations of free Yb$^{3+}$ and Gd$^{3+}$ ions were determined by inductively coupled plasma mass spectrometry (ICP-MS).

### 4.2.5　Cell Cytotoxicity and Cellular Uptake

The cell viability and proliferation of cells were evaluated by the MTT assay. Briefly, A549 cells were incubated in the culture medium at 37 °C in an atmosphere of 5% $CO_2$ and 95% air. After 24 h, the culture medium was replaced with the fresh medium containing different concentrations of PEG-UCNPs for another 24 h. After washing with PBS twice, 100 mL of new culture medium containing MTT reagent (10%) was introduced, followed by further incubation for 4 h to allow formation of formazan dye. After removal of the medium, the purple formazan product was dissolved with DMSO for 15 min. Finally, the optical absorption of formazan at 570 nm was measured by an enzyme-linked immunosorbent assay reader. The upconversion luminescence imaging of cells was carried out using dark field microscope with a 980 nm laser as excitation light source. In vitro cellular uptake of PEG-UCNPs experiment was performed by incubating HeLa cells with different concentrations of PEG-UCNPs (0, 1.6, 3.2 mg Yb/mL) and Iobitridol (10 mg I/mL) for 24 h, followed by washing with PBS. The upconversion luminescence imaging of the cell pellet was performed on Kodak In Vivo Imaging System equipped with a 980 nm laser.

### 4.2.6  CT Imaging

For in vitro CT imaging, PEG-UCNPs and Iobitridol dispersed in water at different Yb or I concentrations including 0, 0.625, 1.25, 2.5, 5, 10, 20, and 30 mg/mL were separately added into the 1.5 mL centrifuge tubes. In addition, $HAuCl_4 \cdot 3H_2O$, $H_2PtCl_6 \cdot 6H_2O$, $BiCl_3$, $YbCl_3$ and tantalum (V) ethoxide were dispersed in ethanol with metal concentration of 10 mg/mL. The CT imaging was performed at 120 KVp.

For in vivo CT imaging, two rats were first anesthetized by intraperitoneal injection of chloral hydrate solution (10 wt%), and then 1 mL of PEG-UCNPs solution (70 mg Yb/mL) and 0.3 mL of Iobitridol (350 mg I/mL) were injected intravenously into two rats through the tail veil, respectively. 50 μl of PEG-UCNPs solution was subcutaneously injected into the rat's paw for lymph-node imaging, and the rat was imaged at indicated time interals. CT images were collected using a JL U.A NO.2 HOSP Philips iCT 256 slice scanner, imaging parameters were as follows: thickness, 0.9 mm; pitch, 0.99; 120 KVp, 300 mA; field of view, 350 mm; gantry rotation time, 0.5 s; table speed, 158.9 mm/s.

### 4.2.7  In Vivo Upconversion Luminescence Imaging

The Kunming mouse was first anesthetized by intraperitoneal injection of chloral hydrate solution (10 wt%), and 100 μl of PEG-UCNPs solution was then subcutaneously injected into the mouse body. The fluorescence imaging was performed on Kodak In Vivo Imaging System equipped with a 980 nm laser.

### 4.2.8  In Vivo MRI

PEG-UCNPs were dispersed in water with Gd concentrations ranging from 0 to 8 mM. To do in vivo MRI, the rat was first anesthetized by intraperitoneal injection of chloral hydrate solution (10 wt%), and then intravenously injected with 1 mL PEG-UCNPs dispersion in water with 10 mM of Gd. The T1-weighted images were acquired using a 1.5 T human clinical scanner.

### 4.2.9  Characterization

TEM images were taken by using a TECNAI G2 high-resolution transmission electron microscope. The molar ratios of rare-earth ions in the products were determined by an ELAN 9000/DRC ICP-MS system. XRD patterns were collected on a D8 ADVANCE (Germany) using Cu K$\alpha$ (0.15406 nm) radiation. XPS measurements were conducted with a VG ESCALAB MKII spectrometer. The XPSPEAK software (Version 4.1) was used to deconvolute the narrow-scan XPS spectra of the C 1 s of the sample. The UC emission spectra were collected on a Photon Technology International (PTI) Timemaster fluorescence spectrometer. FTIR analysis was performed on a Bruker Vertex 70 spectrometer (2 cm$^{-1}$). TGA measurements were carried out by using a Perkin-Elmer TGA-2 thermogravimetric analyzer under nitrogen from room temperature to 800 °C at 10 °C/min. NMR was carried out on a Varian Infinityplus 400 spectrometer operating at a magnetic field strength of 9.4 T.

## 4.3  Results and Discussion

### 4.3.1  Synthesis and Characterization of Yb-Containing UCNPs

Oleic acid (OA)-stabilized UCNPs were synthesized by the strategy as shown in Fig. 4.1a. The as-prepared OA-stabilized UCNPs could be well dispersed in many organic solvents without any detectable aggregation. TEM imaging showed that undoped NaYbF$_4$:Er nanoparticles were monodisperse with irregular shapes and different sizes, whereas the control over size and shape of these nanoparticles was achieved by the introduction of Gd$^{3+}$ ions during the synthetic process. Uniform nanoparticles with regular shapes were produced at a Gd concentration of 20 mol% or higher (Fig. 4.1d, e). This control over size and shape of nanoparticles is very important for biological and clinical applications, as the circulation time in vivo and interactions of the nanoparticles with various cells are highly dependent on their size, shape, and surface properties. High-resolution TEM images revealed lattice

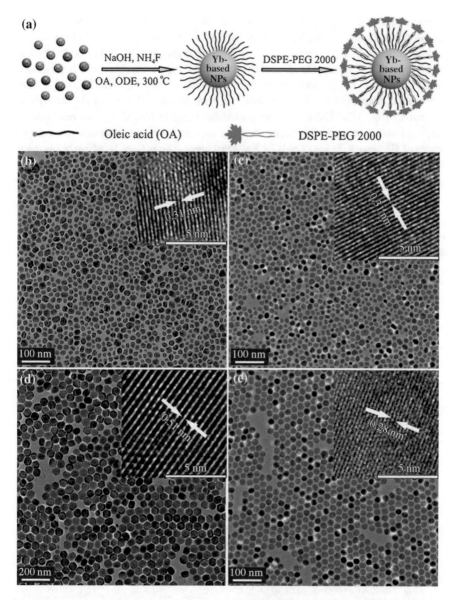

**Fig. 4.1 a** Schematic illustration of the synthesis and surface modification of OA-stabilized UCNPs. Blue $Yb^{3+}$, pink $Gd^{3+}$, and green $Er^{3+}$. TEM and high-resolution TEM images of $NaYbF_4$:Er nanoparticles doped with various concentrations of Gd in chloroform: **b** without doping, **c** 10 mol%, **d** 20 mol% (diameter ca. 52 nm), and **e** 30 mol% (diameter ca. 20 nm). The insets show the corresponding high-resolution TEM images and arrows highlight **d** spacing. (Copyright 2012 Wiley-VCH)

fringes with a spacing of about 0.31 nm for undoped $NaYbF_4$:Er and the 10 mol% Gd-doped samples, which is ascribed to the lattice spacing in the (111) planes of cubic-phase $NaYbF_4$. Interestingly, we found that the spacing of the lattice fringes were 0.51 and 0.28 nm for 20 and 30 mol% Gd-doped samples, which were well consistent with (100) and (101) phases of the hexagonal $NaYbF_4$ crystalline structures, respectively. XRD analysis further confirmed the doping of Gd, as shown in Fig. 4.2a. Cubic sodium rare earth fluoride ($NaREF_4$) has one type of high-symmetry cation site, while hexagonal-phase $NaREF_4$ contains an ordered array of $F^-$ ions with two types of low-symmetry cation sites selectively occupied by $Na^+$ and $RE^{3+}$ ions, leading to significant electron-cloud distortion. As the diameter of $Gd^{3+}$ is larger than that of $Yb^{3+}$, the Gd doping thus allows for a high

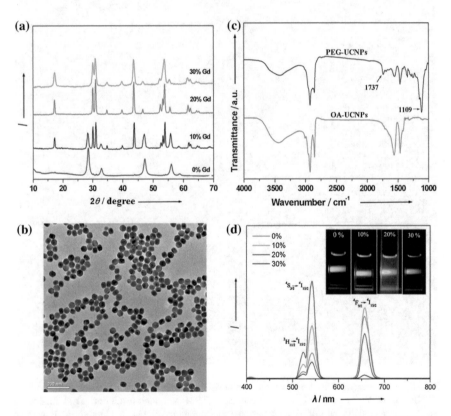

**Fig. 4.2 a** XRD patterns of Gd-doped UCNPs with varied concentrations of Gd (JCPDS NOs. 77-2043 and 27-1427 for cubic and hexagonal $NaYbF_4$ crystalline structure, respectively). **b** TEM image of PEG-UCNPs dispersed in water. **c** FTIR spectra of OA-UCNPs and PEG-UCNPs. **d** Room temperature IR-to-Vis upconversion luminescence spectra of Gd-doped UCNPs with different concentrations of Gd. The electronic states of Er are given next to the corresponding peaks. The inset shows the upconversion luminescence picture of the Gd-doped UCNPs with different concentrations of Gd under excitation at 980 nm. (Copyright 2012 Wiley-VCH)

tendency towards electron-cloud distortion owing to the increased dipole polarizability and thereby facilitates the transformation of the hexagonal structures [11].

As we did not see any peak corresponding to Gd, we thus use EDAX and XPS to examine whether Gd presented in the nanoparticles. The characteristic peaks of Gd were shown in the Fig. 4.3 and Fig. 4.4 [12]. With the ICP-MS analysis, the ratio of Gd in each sample was summarized in Table 4.1.

To make the as-prepared UCNPs water soluble, the OA-UCNPs were modified with DSPE-PEG2000 to yield PEG-UCNPs through the hydrophobic interaction between DSPE-PEG2000 and oleic acid. DSPE-PEG2000 is a well-known biocompatible polymer and has been widely used to stabilize a variety of other nanoparticles for in vivo imaging [13, 14]. The example of 20 mol% Gd-doped OA-UCNPs shows that the as-prepared PEG-UCNPs were well-dispersed in water without any detectable agglomeration (Fig. 4.2b), and they remained stable for days. Even after dialysis for one week in blood serum and physiological saline, no $Yb^{3+}$ or $Gd^{3+}$ leakage were found by ICP-MS. The successful modification of PEG was confirmed by FTIR spectroscopy (Fig. 4.2c). Two new bands at 1737 and 1109 $cm^{-1}$ in the FTIR spectrum of PEG-UCNPs were respectively corresponding to the stretching vibration of the carboxyl ester and the ether bond of PEG chains [15]. The amount of PEG on the surface of nanoparticles was approximately 20%, as confirmed by the thermogravimetric analysis (Fig. 4.5).

## 4.3.2 CT Contrast Efficiency

To evaluate the CT contrast efficacy of PEG-UCNPs, we compared the X-ray absorption of PEG-UCNPs to that of iobitridol, a clinically widely used CT contrast agent in CT imaging. Both agents exhibited signal enhancement as a function of the concentration (Fig. 4.6a, b), and showed a good linear correlation between the Hounsfield units (HU) value and the concentration of Yb (or I). Impressively, at equivalent concentrations, the X-ray absorption of PEG-UCNPs was much higher than that of iobitridol (Fig. 4.6a). Considering that the high-Z metal elements contribute to the dominant X-ray attenuation, we also examined, as a proof-of-concept study, the X-ray attenuation of several metallic salts to better illustrate the contrast efficiency of PEG-UCNPs compared to currently available Au-, Pt-, Bi-, and Ta-based nanoparticulate CT contrast agents (Fig. 4.7a). The result showed that Yb exhibited the highest HU value at 120 kVp, a commonly used voltage in clinical applications. As we discussed before, the K-edge of Yb (61 keV) is located just within the highest engery region of the X-ray spectrum at this voltage, whereas the K-edge values of I (33 keV), Ta (67 keV), Pt (78 keV), Au (81 keV), and Bi (91 keV) were deviated from this region (Fig. 4.7b). These results strongly suggested the great potential of the as-prepared PEG-UCNPs as highly efficient CT contrast agents at 120 KVp.

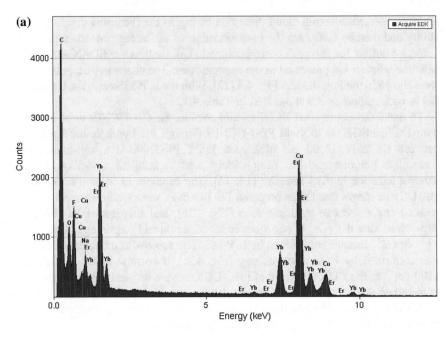

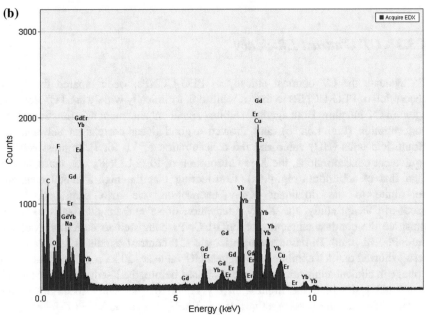

**Fig. 4.3** EDAX spectra of NaYbF$_4$:2%Er (**a**) and NaYbF$_4$:20%Gd,2%Er (**b**). (Copyright 2012 Wiley-VCH)

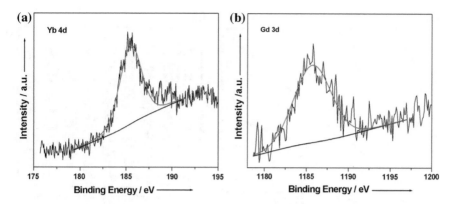

**Fig. 4.4** Gd 3d XPS spectra of NaYbF₄:2%Er (**a**) and NaYbF₄:20%Gd,2%Er (**b**). (Copyright 2012 Wiley-VCH)

**Table 4.1** The molar ratio of different elements in each sample. (Copyright 2012 Wiley-VCH)

| Samples | NaYbF₄:Er | NaYbF₄:Gd,Er (10% Gd) | NaYbF₄:Gd,Er (20% Gd) | NaYbF₄:Gd,Er (30% Gd) |
|---|---|---|---|---|
| Stoichiometric molar ratio of Y/Yb/Er/Tm | 0.98:0.02 | 0.88:0.10:0.02 | 0.78:0.20:0.02 | 0.68:0.30:0.02 |
| Actual molar ratio of Y/Yb/Er/Tm | 0.982:0.018 | 0.90:0.082:0.018 | 0.79:0.191:0.019 | 0.678:0.306:0.02 |

**Fig. 4.5** TGA curves of NaYbF₄:20%Gd,2%Er before (**a**) and after (**b**) PEG modification. (Copyright 2012 Wiley-VCH)

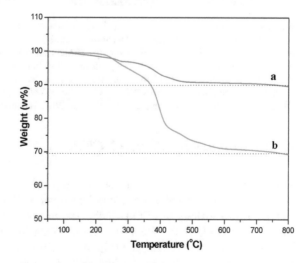

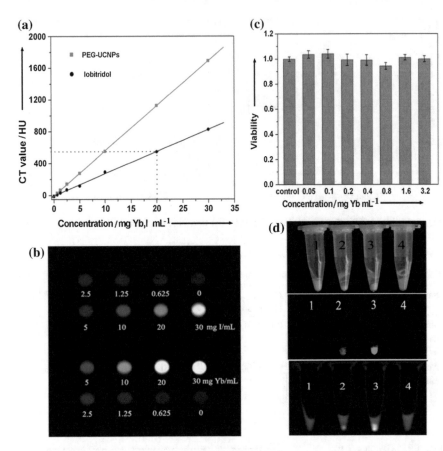

**Fig. 4.6** **a** HU values and **b** CT images of PEG-UCNPs (red squares) and iobitridol (black circles) as a function of the concentration of Yb (red trace) and I (black trace), respectively. **c** Cell viability of A549 cells after incubation with increased concentration of PEG-UCNPs for 24 (h. d) Cell uptake analysis: bright field images (top), upconversion luminescence images (middle), and CT images (bottom) of washed pellets from HeLa cells incubated without contrast agent (1), with different concentrations of PEG-UCNPs (1.6 and 3.2 mgYb/mL for (2) and (3), respectively), and iobitridol (10 mgI/mL for (4)) for 24 h. (Copyright 2012 Wiley-VCH)

## 4.3.3   Cytotoxicity of PEG-UCNPs

Prior to in vivo imaging, we first examined the toxicity of PEG-UCNPs. Figure 4.8 demonstrated that at tested concentrations, over 95% cells remained alive, indicating the low toxicity of PEG-UCNPs. The low toxicity may be due to the good stability of PEG-UCNPs. We next examined their imaging in vitro. As seen from the fluorescence and CT images (Fig. 4.6d), PEG-UCNPs were effectively internalized by cancer cells and the internalization was concentration-dependent. In sharp contrast, cells showed little internalization of iobitridol, even at a

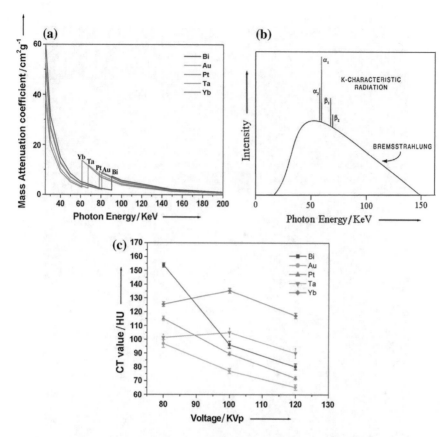

**Fig. 4.7** **a** Mass attenuation coefficients of different metal elements as a function of X-ray energy. **b** Typical X-ray emission spectrum from a tungsten X-ray tube operating at 150 kVp with aluminum filtering. **c** CT values of different metallic salts dispersed in ethanol with metal concentration of 10 mg/mL at varied voltages. (Copyright 2012 Wiley-VCH)

concentration of 10 mg/mL, which was three times higher than that of Yb in PEG-UCNPs. The internalization of PEG-UCNPs by cancer cells is of great significance in clinical applications, in particular for tumor target imaging and detection in vivo (Table 4.2).

## 4.3.4 In Vivo CT Imaging of PEG-UCNPs

Given the promising in vitro results, we next evaluated the in vivo CT imaging performance of PEG-UCNPs. After intravenous injection of PEG-UCNPs, the whole-body CT scanning of the rat was performed. As a reference, iobitridol was also injected into another rat for whole-body CT scanning. As shown in Fig. 4.8, a

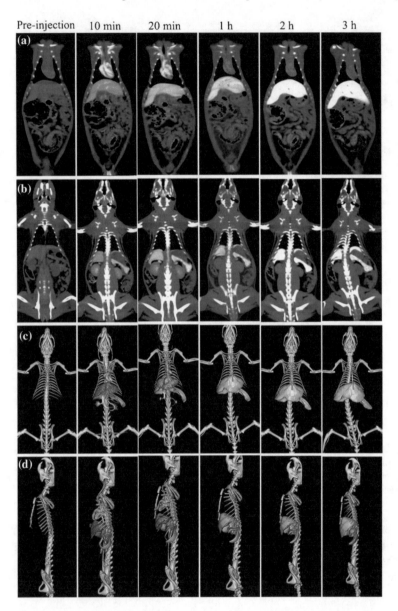

**Fig. 4.8** In vivo CT coronal view images of a rat after intravenous injection of 1 mL of PEG-UCNPs (70 mgYb/mL) solution at timed intervals. **a** Heart and liver. **b** Spleen and kidney. **c**, **d** The corresponding 3D renderings of in vivo CT images. (Copyright 2012 Wiley-VCH)

clear enhancement of the signal in the heart was observed at least within 20 min postinjection without an obvious contrast loss, and a more careful look at the 3D-renderings of CT images revealed evident enhancement of the signals of great

**Table 4.2** CT values of different organs in the rat after injection of PEG-UCNPs (1 mL, 75 mg Yb/mL). (Copyright 2012 Wiley-VCH)

| Time | Heart | Liver | Spleen | Kidney |
|------|-------|-------|--------|--------|
| Pre-injection | 56.1 | 80.2 | 76.3 | 45.5 |
| 10 min | 391.4 | 191.2 | 252.8 | 129.3 |
| 20 min | 336.5 | 268.1 | 294.3 | 106.0, 237.6 |
| 1 h | 138.1 | 451.7 | 409.5 | 83.4 |
| 2 h | 108.8 | 561.9 | 502.9 | 83.5 |
| 3 h | 61.3 | 520.9 | 344.5 | 62.9 |

vessels within the same period of time. Long-term circulation in vessels is critical in many biomedical applications, including vascular imaging and diagnosis of diseases. After one hour, the signals of the heart and vessels decreased rapidly, while the gradual enhancement of the signals for liver and spleen continued for over two hours. This long-lasting liver-signal enhancement may improve the detection of the hepatic metastases [16]. More impressively, a good contrast effect was also obtained when the injection dose was decreased by half (Fig. 4.9 and Table 4.3). The reduced dose requirement is highly beneficial in a contrast agent, because it will minimize the potential adverse effects associated with the injection of contrast agents in patients.

In contrast, the commercial CT contrast agent was quickly cleared from the blood circulation, and no contrast signal was seen in the heart, liver, and spleen. Most contrast agent accumulated in the kidney and bladder (Fig. 4.10 and Table 4.4). Such a short circulation time make it hard for lymph node mapping to improve cancer staging, which would be helpful for avoiding unnecessary surgery to stage the cancer [17]. On the contrary, the long circulation of PEG-UCNPs in vessels, as illustrated above, would enable the nanoparticles to migrate to the lymph nodes through lymphatic drainage, as confirmed in Fig. 4.11.

### 4.3.5 In Vivo Side Effects of PEG-UCNPs

To further determine whether PEG-UCNPs can trigger any harmful effects, the rat were intravenously injected with PEG-UCNPs, and the organs including heart, liver, and spleen were harvest after three weeks for hematoxylin and eosin (H&E) staining. No tissue damage or any other adverse effect associated with the administration of PEG UCNPs was observed (Fig. 4.12). This result suggested that PEG-UCNPs hold great promise as an injectable CT contrast agent for applications in biological medicine. Notably, about 34% of the injected PEG-UCNPs were cleared from the rat body during this period of time, as confirmed by ICP-MS analysis. The relatively slow elimination has also been observed in other nanoparticulate imaging agents [18], and the elimination route and long-term toxicity need further studies.

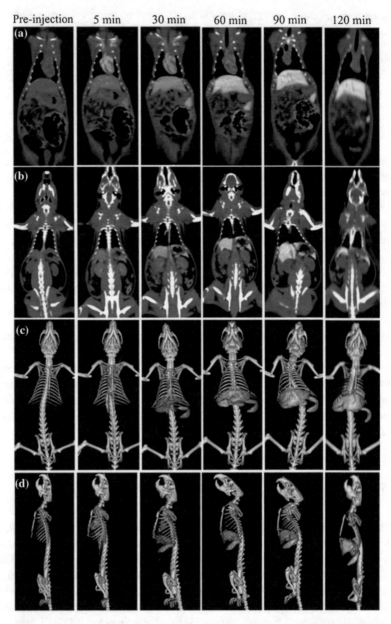

**Fig. 4.9** In vivo CT coronal view images of a rat after intravenous injection of half dose of PEG-UCNPs in Fig. 4.8. **a** Heart and liver. **b** Spleen and kidney. **c, d** The corresponding 3D renderings of in vivo CT images. (Copyright 2012 Wiley-VCH)

**Table 4.3** CT values of different organs in the rat after injection of PEG-UCNPs (0.5 mL, 75 mg Yb/mL). (Copyright 2012 Wiley-VCH)

| Time | Heart | Liver | Spleen | Kidney |
|---|---|---|---|---|
| Pre-injection | 51.4 | 61.8 | 58.3 | 48.2 |
| 5 min | 195.9 | 119.4 | 134.1 | 86.5 |
| 30 min | 161.2 | 182.2 | 201.5 | 70.6 |
| 1 h | 89.8 | 298.7 | 255.9 | 54.0 |
| 2 h | 62.9 | 344.5 | 284.4 | 49.5 |
| 24 h | 63.2 | 291.5 | 312.9 | 32.7 |

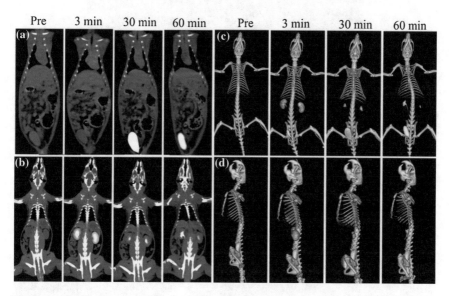

**Fig. 4.10** In vivo CT coronal view images of a rat after intravenous injection of 0.3 mL of iobitridol (305 mg I/mL). **a** Heart and liver. **b** Spleen and kidney. **c, d** The corresponding 3D renderings of in vivo CT images. (Copyright 2012 Wiley-VCH)

**Table 4.4** CT values of different organs in the rat after injection of iobitridol (305 mg I/mL). (Copyright 2012 Wiley-VCH)

| Time | Heart | Liver | Kidney |
|---|---|---|---|
| Pre-injection | 65.3 | 80.5 | 59.5 |
| 3 min | 111.1 | 87.4 | 669.0 |
| 30 min | 64.0 | 68.8 | 599.9,63.1 |
| 1 h | 66.8 | 80.6 | 703.8,64.6 |
| 24 h | 64.8 | 59.2 | 37.4 |

## 4.3.6   Multi-Modal Imaging

$NaYbF_4$ is a common host for upconversion fluorescence materials. It is known that upconversion fluorescence is superior to down-conversion fluorescence for biomedical applications, as it emits high-energy lights under excitation with the low-energy light (NIR). In addition, human tissues have low absorption of the NIR

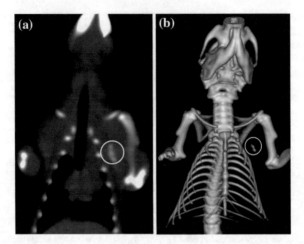

**Fig. 4.11** **a** CT images of a lymph node of a rat at 60 min after subcutaneous administration of 50 μL PEG-UCNPs solution into the paw. **b** The corresponding 3D-volume renderings of the above CT image. (Copyright 2012 Wiley-VCH)

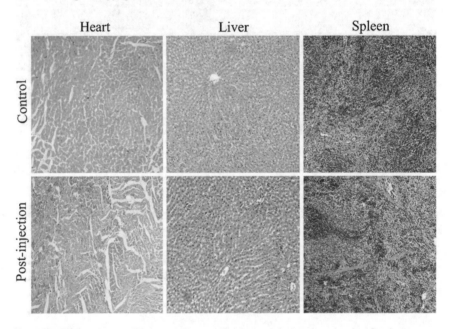

**Fig. 4.12** H&E staining of the heart, liver, and spleen of the rat three weeks after intravenous injection of PEG-UCNPs. (Copyright 2012 Wiley-VCH)

light, thus simultaneously minimizing the phototoxicity and background signals [19, 20]. Interestingly, the Gd doping induced drastically enhanced fluorescence compared to undoped NPs. We hypothesized that the fluorescence enhancement was a result of Gd-doping induced cubic-to-hexagonal phase conversion and the

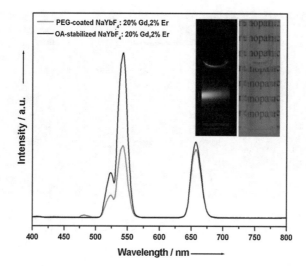

**Fig. 4.13** Room-temperature infrared-to-visible upconversion luminescence spectrum of OA-stabilized NaYbF$_4$:20%Gd,2%Er nanoparticles in chloroform and PEG-coated NaYbF$_4$:20% Gd,2%Er nanoparticles in water under excitation at 980 nm. The inset shows the corresponding luminescence and bright-field photos of PEG-coated NaYbF$_4$:20%Gd,2%Er nanoparticles in water. (Copyright 2012 Wiley-VCH)

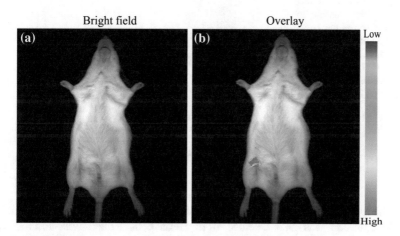

**Fig. 4.14 a** White light photograph and **b** in vivo upconversion luminescence image of a mouse after subcutaneous administration of PEG-UCNPs. (Copyright 2012 Wiley-VCH)

size increase of nanocrystals [21, 22]. After PEGylation, no obvious change in the upconversion fluorescence wavelength except a slight decrease in fluorescence intensity, which is in large due to the change of ligand and solvent (Fig. 4.13). Next, 100 µL of PEG-UCNPs dispersion in water was subcutaneously injected into a Kunming mouse to assess their in vivo fluorescence imaging. As seen from

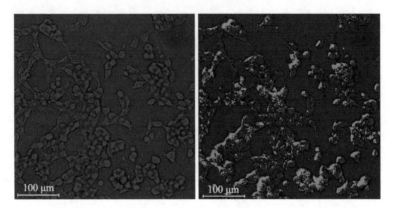

**Fig. 4.15** Bright field and luminescence images under the excitation at 980 nm of HeLa cells incubated with PEG-UCNPs solution with a Yb concentration of 3.2 mg/mL at 37 °C for 4 h. (Copyright 2012 Wiley-VCH)

**Fig. 4.16** Room-temperature infrared-to-visible upconversion luminescence spectrum of OA-stabilized NaYbF₄:20%Gd,2%Tm nanoparticles in chloroform under excitation at 980 nm. The inset shows the corresponding luminescence photo of PEG-coated NaYbF4:20%Gd,2%Tm nanoparticles in chloroform. (Copyright 2012 Wiley-VCH)

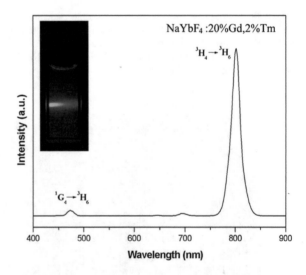

Fig. 4.14, a bright emission was clearly noticed at the injection site. Moreover, this bright emission was also detected from HeLa cells treated with PEG-UCNPs (Fig. 4.15). Notably, the emission could be further tuned to the NIR region (800 nm) (Fig. 4.16), which will help to further increase the tissue penetration and thus improve the detection sensitivity [23].

In addition to upconversion fluorescence, Gd doping can further endow the NPs with MRI capability. We therefore evaluated the MRI capability of PEG-UCNPs. Figure 4.17 is the T1-weighted MR image of the PEG-UCNPs at different Gd concentrations. The MR signal was enhanced linearly as the Gd concentration increased over the range from 0 to 8 mM. The specific relaxivity values (r1) calculated from the slope of the concentration-dependent relaxation rate was

**Fig. 4.17** **a** The linear relationship between T1 relaxation rates (1/T1) and Gd concentrations **b** T1-weighted magnetic resonance image of PEG-UCNPs dispersed in water with different Gd concentrations from a 1.5 T clinical MRI system. **c** The leaching experiment of $Gd^{3+}$ from PEG-UCNPs matrix. Arsenazo III can be form arsenazo-$Gd^{3+}$ complex with free $Gd^{3+}$ ions, inducing a new absorption peak at 658 nm. (Copyright 2012 Wiley-VCH)

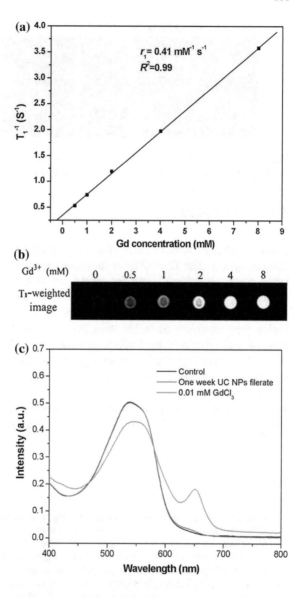

$0.41 \text{ mM}^{-1} \text{ s}^{-1}$, which is comparable to the values of previous Gd-based and MnO T1 contrast agents [24–27]. Significantly, no toxic free $Gd^{3+}$ ion leached from the PEG-UCNPs even after one week (Fig. 4.17c), thus indicating that PEG-UCNPs were stable and capable of use for in vivo imaging without potential $Gd^{3+}$ poisoning. To assess their in vivo imaging effects, 1 mL of PEG-UCNPs dispersion at a Gd concentration of 10 mM was intravenously injected into a rat and the rat was imaged by a 1.5 T human clinical scanner. T1-weighted enhancement was clearly visible as shown in Fig. 4.18.

Pre-injection                          Post-injection

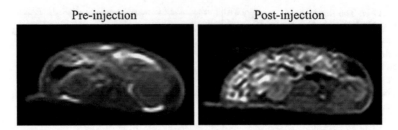

**Fig. 4.18** In vivo MRI images of a rat before vs after intravenous injection of PEGUCNPs solution with the Gd concentration of 10 mM. (Copyright 2012 Wiley-VCH)

## 4.4    Conclusion

In this chapter, we have described the first example of Yb-based nanoparticulate CT contrast agents. The feasibility of these NPs as an in vivo CT contrast agent was systemically investigated, and the results revealed that PEG-UCNPs showed low cytotoxicity and long circulation time in vivo. More significantly, the contrast efficacy is significantly enhanced compared to a clinical iodinate agent, and even higher than currently available Au-, Pt-, Bi-, and Ta-based nanoparticulate CT contrast agents under normal operating conditions (120 kVp). This improvement is primarily attributed to the K-edge energy of Yb which is well match with the X-ray spectrum. Furthermore, Gd-doping in these nanoparticles allows these NPs for enhanced fluorescence and NMR imaging. This multimodal imaging property would allow these nanoparticles to provide a better reliability of the collected data and have great potential in biological and medical applications.

## References

1. Kalender WA (2006) X-ray computed tomography. Phys Med Biol 51:R29–R43
2. deKrafft KE, Xie Z, Cao G et al (2009) Iodinated nanoscale coordination polymers as potential contrast agents for computed tomography. Angew Chem Int Ed 48:9901–9904
3. Hyafil F, Cornily JC, Feig JE et al (2007) Noninvasive detection of macrophages using a nanoparticulate contrast agent for computed tomography. Nat Med 13:636–641
4. Ai K, Liu Y, Liu J et al (2011) Large-scale synthesis of $Bi_2S_3$ nanodots as a contrast agent for in vivo x-ray computed tomography imaging. Adv Mater 23:4886–4891
5. Yu SB, Watson AD (1999) Metal-based x-ray contrast media. Chem Rev 99:2353–2377
6. Cheung ENM, Alvares RDA, Oakden W et al (2010) Polymer-stabilized lanthanide fluoride nanoparticle aggregates as contrast agents for magnetic resonance imaging and computed tomography. Chem Mater 22:4728–4739
7. Hyafil F, Cornily JC, Feig JE et al (2007) Noninvasive detection of macrophages using a nanoparticulate contrast agent for computed tomography. Nat Med 13:636–641
8. Krause W, Schuhmann-Giampieri G, Bauer M et al (1996) Dysprosium-EOB-DTPA. A new prototype of liver-specific contrast agents for computed tomography. Invest Radiol 31:502–511

9. Schmitz SA, Wagner S, Schuhmann-Giampieri G et al (1997) Gd-EOB-DTPA and Yb-EOB-DTPA: two prototypic contrast media for CT detection of liver lesions in dogs. Radiology 205:361–366

10. Xiong LQ, Yang TS, Yang Y et al (2010) Long-term in vivo biodistribution imaging and toxicity of polyacrylic acid-coated upconversion nanophosphors. Biomaterials 31:7078–7085

11. Wang F, Han Y, Lim CS et al (2010) Simultaneous phase and size control of upconversion nanocrystals through lanthanide doping. Nature 463:1061–1065

12. Yang HS, Santra S, Walter GA et al (2006) $Gd^{III}$-Functionalized fluorescent quantum dots as multimodal imaging probes. Adv Mater 18:2890–2894

13. Dubertret B, Skourides P, Norris DJ et al (2002) In vivo imaging of quantum dots encapsulated in phospholipid micelles. Science 298:1759–1762

14. Jin YD, Jia CX, Huang SW et al (2010) Multifunctional nanoparticles as coupled contrast agents. Nat Commun. doi:10.1038/ncomms1042

15. Liu D, Wu W, Ling J et al (2011) Effective PEGylation of iron oxide nanoparticles for high performance in vivo cancer imaging. Adv Funct Mater 21:1498–1504

16. Kim D, Park S, Lee JH et al (2007) Gold nanoparticles as high-resolution X-ray imaging contrast agents for the analysis of tumor-related micro-vasculature. J Am Chem Soc 129:7661–7665

17. Harisinghani MG, Barentsz J, Hahn PF et al (2003) Noninvasive detection of clinically occult lymph-node metastases in prostate cancer. J Med 348:2491–2499

18. Rabin O, Perez JM, Grimm J et al (2006) An X-ray computed tomography imaging agent based on long-circulating bismuth sulphide nanoparticles. Nat Mater 5:118–122

19. Zhan Q, Qian J, Liang H et al (2011) Using 915 nm laser excited $Tm^{3+}/Er^{3+}/Ho^{3+}$-doped $NaYbF_4$ upconversion nanoparticles for in vitro and deeper in vivo bioimaging without overheating irradiation. ACS Nano 5:3744–3757

20. Ehlert O, Thomann R, Darbandi M et al (2008) A four-color colloidal multiplexing nanoparticle system. ACS Nano 2:120–124

21. Chen G, Ohulchanskyy TY, Kumar R et al (2010) Ultrasmall monodisperse $NaYF_4$:$Yb^{3+}$/$Tm^{3+}$ nanocrystals with enhanced near-infrared to near-infrared upconversion photoluminescence. ACS Nano 4:3163–3168

22. Mai HX, Zhang YW, Si R et al (2006) High-quality sodium rare-earth fluoride nanocrystals: Controlled synthesis and optical properties. J Am Chem Soc 128:6426–6436

23. Baumes JM, Gassensmith JJ, Giblin J et al (2010) Storable, thermally activated, near-infrared chemiluminescent dyes and dye-stained microparticles for optical imaging. Nat Chem 2:1025–1030

24. Kumar R, Nyk M, Ohulchanskyy TY et al (2009) Combined optical and MR bioimaging using rare earth ion doped $NaYF_4$ nanocrystals. Adv Funct Mater 19:853–859

25. Na HB, Lee JH, An K et al (2007) Development of a T1 contrast agent for magnetic resonance imaging using mno nanoparticles. Angew Chem Int Ed 46:5397–5401

26. Zhou J, Yu MX, Sun Y et al (2011) Fluorine-18-labeled $Gd^{3+}/Yb^{3+}/Er^{3+}$ co-doped $NaYF_4$ nanophosphors for multimodality PET/MR/UCL imaging. Biomaterials 32:1148–1156

27. Boyer JC, Vetrone F, Cuccia LA et al (2006) Synthesis of colloidal upconverting $NaYF_4$ nanocrystals doped with $Er^{3+}$, $Yb^{3+}$ and $Tm^{3+}$, $Yb^{3+}$ via thermal decomposition of lanthanide trifluoroacetate precursors. J Am Chem Soc 128:7444–7445

# Chapter 5
# Hybrid BaYbF₅ Nanoparticles: Novel Binary Contrast Agent for High-Resolution in Vivo X-Ray Computed Tomography Angiography

**Abstract** In clinical CT examination, the operating voltage changes from 80 to 140 KVp depending on the applications. All the contrast agents that contain single contrast element can only provide limited contrast efficacy and cannot be tailored to the changes of operating voltage. To overcome this hurdle, we describe, in this chapter, a binary nanoparticulate CT contrast agent by integrating Yb and Ba into a single nanoparticle. Owing to the big difference between the K-edge energy of these elements, this binary contrast agent can provide much higher contrast efficiency than that of the clinical iodinated contrast agent, regardless of the voltage setting. More importantly, this agent has low toxicity and further improved in vivo imaging effect, which is successfully used for in vivo X-ray computed tomography angiography.

## 5.1 Introduction

In previous chapter, we have discussed that the difference X-ray attenuation capabilities among different elements derives from the difference in the K-edge energies. When the operating voltage is 120 keV, the UCNPs have shown significantly enhanced contrast due to the suitable K-edge energy of Yb, which is well math with the X-ray spectrum. However, the operating voltage is generally changed individually. For example, the operating voltage will be decreased to 80 KVp when the patients are children to minimize any potential radiation effects, whereas it will be increased to 140 KVp for overweighted patients to get more accurate diagnostic information. We found that there is only one contrast element for all currently studied nanoparticle-based CT contrast agents such as Au in gold NPs, Bi in $Bi_2S_3$ NPs, Ta in TaOx NPs, and I in iodine-containing small molecule-based CT contrast agents. The single contrast element can not response to the change of the operating voltage to generate sufficient contrast at all the operating voltages, including the Yb-based UCNPs in previous chapter. At a given voltage, the contrast efficiency of the aforementioned heavy element-containing NPs might be lower than that of the commercial iodine-based small molecules. In addition, the single contrast element

© Springer Nature Singapore Pte Ltd. 2018
Y. Liu, *Multifunctional Nanoprobes*, Springer Theses,
DOI 10.1007/978-981-10-6168-4_5

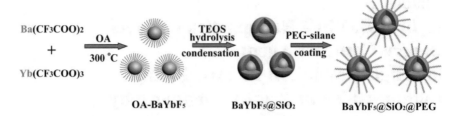

**Fig. 5.1** Schematic illustration of the synthesis of BaYbF$_5$@SiO$_2$@PEG. (Copyright 2012 Wiley-VCH)

can only provide limited contrast effect and cannot be tailored to the changes of operating voltage. The construction of nanoparticulate CT contrast agents consisting of multiple elements that have differential K-edge values within the X-ray spectrum of CT would be an promising strategy to achieve high CT contrast at all voltages. Unfortunately, to our best knowledge, those CT contrast agents have remained elusive.

In clinical applications, barium sulfate has been widely for diagnosis of intestine diseases. Similar to I (31 keV), the K-edge energy of Ba is 37 keV. Thus, Ba has high contrast efficiency at lower operating voltage, similar to the case of I. As we discussed in previous chapter, Yb will produce strong X-ray attenuation at higher operating voltage. Thus, we hypothesized that the combination of Yb and Ba in a single nanoparticle will produce high contrast at different operating voltages.

In this chapter, we reported a new generation of CT contrast agent that contains Yb and Ba in a single nanoparticle denoted hereafter as BaYbF$_5$@SiO$_2$@PEG (Fig. 5.1). This binary contrast agent not only offered much higher CT contrast efficiency than clinical iodinated agents and currently studied nanoparticulate CT contrast agents which are composed of a single contrast element, but also maintained high X-ray attenuation at different voltages for potential diagnostic different patients. Moreover, surface modification of anti-biofouling PEG-silane allows the nanoparticles to have prolonged circulation time in vivo and low toxicity, and thereby they could be used for blood pool CT imaging.

## 5.2   Experimental Section

### 5.2.1   Materials

The rare-earth oxides Yb$_2$O$_3$ (purity > 99.99%) was obtained from Changchun Hepalink rare-earth materials company. Oleic acid and 1-octadecene were purchased from Sigma-Aldrich. Iobitridol was purchased from Guerbet (France).

Ba(OH)$_2$·8H$_2$O was purchased from Alfa. Fluorescein isothiocyanate (FITC), 3-aminopropyltriethoxysilane (APTES), tetraethoxysilane (TEOS) were purchased form Aldrich. 2- methoxy(polyethyleneoxy)propyltrimethoxysilane (PEG-silane) was purchased from Meryer Chemical Technology Co. Ltd., (China). All chemicals were analytical grade and used as received without further purification.

## 5.2.2 Synthesis of OA-Stabilized BaYbF$_5$ Nanoparticles

OA-stabilized BaYbF$_5$ nanoparticles were prepared according to a previously reported method with some modification [1]. Briefly, Yb$_2$O$_3$ (1 mmol) and Ba(OH)$_2$ 8H$_2$O (2 mmol) were dissolved in 50% aqueous trifluoroacetic acid at 90 °C. After removal of the water and acid under vacuum at 60 °C, oleic acid (30 mL) was added. The solution was slowly heated to 110 °C under Ar protection and kept at this temperature for 30 min to remove the residual water and oxygen. Subsequently, the resulting yellow solution was heated to 300 °C at a rate of 7 °C/min and remained at this temperature for 1 h. After cooling down to room temperature naturally, OA-stabilized BaYbF$_5$ nanoparticles were precipitated by the addition of acetone and isolated through centrifugation. The precipitate was washed with ethanol several times and redispersed in cyclohexane for further modification.

## 5.2.3 Silica-Coated BaYbF$_5$ Nanoparticles and Surface PEGylation

Silica coated BaYbF$_5$ nanocrystals were prepared through a modified microemulsion method [2]. Briefly, 0.9 ml of Igepal CO-520 was mixed with 18 mL cyclohexane. After ultrasonic treatment for 20 min, 1 mL BaYbF$_5$ solution prepared by diluting the BaYbF$_5$ stock solution (60 μL, 0.15 M) with cyclohexane was then added. After 4 h vigorous stirring at room temperature, 160 μL of ammonia and 80 μL of TEOS were added and the mixture was aged for 24 h for hydrolysis and condensation of the silica precursor. Silica- coated BaYbF$_5$ nanoparticles were isolated by centrifugation at 9500 rpm for 15 min, washed with ethanol and water (1:1) three times, and finally redispersed in isopropanol. Surface PEGylation was performed through the hydrolysis of 2-[methoxy-(polyethyleneoxy)propyl]trimethoxysilane (50 μL) in the mixture solution at 80 °C for 12 h. After isolation by centrifugation and washing with deionized water twice, the obtained BaYbF$_5$@SiO$_2$@PEG NPs were finally dispersed in physiological saline for further characterization.

### 5.2.4   Synthesis of BaYbF$_5$(FITC)@SiO$_2$@PEG NPs

FITC (5 mg) was conjugated to APTES (10 μL) in anhydrous ethanol (3 mL) by an addition reaction between isothiocyanate group in FITC with amine group in APTES. The reaction was allowed to proceed in the dark for 4 h under general stirring. The silica coating and surface PEGylation were carried out under the same procedures as discussed above.

### 5.2.5   In Vitro Toxicity Studies of BaYbF$_5$@SiO$_2$@PEG

In vitro cytotoxicity of BaYbF$_5$@SiO$_2$@PEG was evaluated by the MTT assay. HeLa cells and human embryonic kidney 293 (HEK 293) cells were respectively incubated in the culture medium under standard culture conditions (37 °C, 5% CO$_2$) for 24 h. The medium was then replaced with fresh medium containing different concentrations of BaYbF$_5$@SiO$_2$@PEG ranging from 0 to 12.8 mM for another 24 h. After washing with PBS twice, the cell viabilities was tested using MTT assay.

### 5.2.6   In Vitro and in Vivo CT Imaging

For in vitro CT imaging: BaYbF$_5$@SiO$_2$@PEG, NaYbF$_4$@PEG, and Iobitridol were respectively dispersed in water at concentrations ranging from 0 to 80 mM.

For in vivo CT imaging, the rats were first anesthetized by intraperitoneal injection of chloral hydrate solution (10 wt%), and then intravenously injected with either BaYbF$_5$@SiO$_2$@PEG dispersion in physiological saline (800 μL, 0.2 M) or of Iobitridol (0.5 mL, 350 mg I/mL). For blood pool imaging of the rabbit, the rabbit was first anesthetized by intraperitoneal injection of chloral hydrate solution (10 wt%) and 7 mL BaYbF$_5$@SiO$_2$@PEG dispersion in PBS was intravenously injected into the rabbit via the rabbit's auricular vein. CT images were collected using a JL U.A NO.2 HOSP Philips iCT 256 slice scanner, imaging parameters were as follows: thickness, 0.9 mm; pitch, 0.99; 120 KVp, 300 mA; field of view, 350 mm; gantry rotation time, 0.5 s; table speed, 158.9 mm/s. The effect of the operating voltage on the X-ray attenuation of BaYbF$_5$@SiO$_2$@PEG was investigated at 80–140 KVp (the range of scanning voltage set in clinical CT scanner).

### 5.2.7  Biodistribution of BaYbF$_5$@SiO$_2$@PEG

BaYbF$_5$(FITC)@SiO$_2$@PEG solution was administrated intravenously into a rat. After 24 h, the rat was sacrificed and the distribution of BaYbF$_5$(FITC) @SiO$_2$@PEG in several organs including heart, liver, spleen, kidney, and intestine was monitored by fluorescence imaging and ICP-MS analysis. The fluorescence imaging was performed on Kodak In Vivo Imaging System. For ICP-MS analysis, these organs were lyophilized and weighed and digested in concentrated aqueous HNO$_3$. The concentration of nanoparticles was quantified by ICP-MS.

Animal care and handing procedures were in agreement with the guidelines of the Regional Ethics Committee for Animal Experiments.

### 5.2.8  Characterization

Transmission electron microscopy (TEM) imaging was performed on a TECNAI G2 high-resolution transmission electron microscope with a tungsten filament at an accelerating voltage of 200 kV. The molar ratios of Ba/Yb in the products were determined by an ELAN 9000/DRC ICP-MS system. Powder X-ray diffraction analysis of the as-prepared samples were carried out on a D8 ADVANCE diffractometer (Germany) using Cu Kα (0.15406 nm) radiation. X-Ray photoemission spectroscopy (XPS) measurements were conducted with a VG ESCALAB MKII 250 spectrometer. The XPSPEAK software (Version 4.1) was used to deconvolute the narrow-scan XPS spectra of the C 1 s of the sample. The 29Si NMR spectrum was recorded on a Bruker AVANCE III 400 WB spectrometer equipped with a 4 mm standard bore CP MAS probehead whose X channel was tuned to 79.50 MHz for 29Si and the other channel was tuned to 400.18 MHz for broad band 1H decoupling, using a magnetic field of 9.39 T at 297 K. Fluorescence spectra of FITC-labeled BaYbF$_5$@SiO$_2$@PEG NPs was obtained using a Perkin-Elmer LS 55 luminescence spectrometer.

## 5.3  Results and Discussion

### 5.3.1  Synthesis and Characterization

BaYbF$_5$ nanocrystals were synthesized through a modified thermal decomposition process of their corresponding trifluoroacetate precursors in OA. The as-prepared OA-stabilized BaYbF$_5$ nanocrystals could be well dispersed in many low polarity solvents (e.g., toluene and chloroform) owing to the presence of the capping ligand (OA) on the surface and remained colloidally stable for over one month without obvious aggregation. Transmission electron microscopy (TEM) image showed that the as-prepared OA-BaYbF$_5$ nanocrystals were spherical in shape with an average diameter of about 10 nm and a narrow size distribution (Fig. 5.2a). High-resolution TEM image showed lattice fringes with an observed d-spacing of 0.29 nm, corresponding to the lattice spacing in the (330) phases of the cubic BaYF$_5$ crystalline structures. XRD analysis further confirmed that the peak positions and intensity matched well with the calculated pattern for cubic BaYF$_5$ crystals (JCPDS NO. 46–0039) (Fig. 5.2c). EDAX (Fig. 5.3) provided a strong correlation of these elements with OA-BaYbF$_5$ nanocrystals. The actual molar ratio of Ba/Yb in the final products was closed to 1:1, as determined by inductively coupled plasma mass spectrometry (ICP-MS) (Fig. 5.4).

To enable OA-stabilized BaYbF$_5$ nanocrystals for biomedical applications, OA-stabilized BaYbF$_5$ nanocrystals were coated with silica to make them water soluble [3–6] Silica coating has many merits including (i) facile process and controllable shell sickness; (ii) facile surface modification on the surface; and

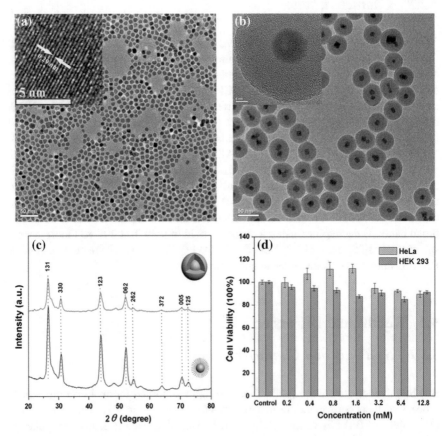

**Fig. 5.2** **a** TEM image of OA-stabilized BaYbF$_5$ nanocrystals. The insert is the corresponding high-resolution TEM image. **b** TEM image of BaYbF$_5$@SiO$_2$@PEG. **c** XRD spectra of the OA-BaYbF$_5$ and BaYbF$_5$@SiO$_2$@PEG. **d** Cell viabilities of HEK 293 and HeLa cells after incubation with increased concentration of BaYbF$_5$@SiO$_2$@PEG for 24 h. (Copyright 2012 Wiley-VCH)

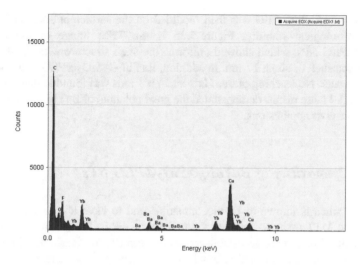

**Fig. 5.3** EDAX spectrum of OA-BaYbF$_5$. (Copyright 2012 Wiley-VCH)

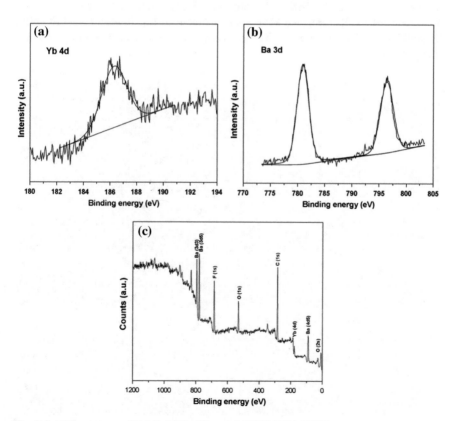

**Fig. 5.4** XPS spectrum of OA-BaYbF$_5$. (Copyright 2012 Wiley-VCH)

(iii) biocompatibility. PEG was then modified on the surface of silica to improve their physiological stability. Figure 5.2b is the TEM image of the BaYbF$_5$ @SiO$_2$@PEG NPs, which showed uniform core-shell structure with the thickness of the silica shell of about 17 nm. In addition, BaYbF$_5$@SiO$_2$@PEG NPs remained stable in water. No leaching of free Ba$^{2+}$ and Yb$^{3+}$ ions was found as determined by ICP-MS. All these results demonstrated the great potential of BaYbF$_5$@SiO$_2$@PEG NPs for in vivo applications.

### 5.3.2   Cytotoxicity of BaYbF$_5$@SiO$_2$@PEG NPs

Although silica is known to be biocompatible and to have low toxicity, we performed the MTT assay to evaluate the toxicity of the BaYbF$_5$@SiO$_2$@PEG NPs. This crucial factor must be established in determining its feasibility for specific applications in experimental animal studies. Herein, the effect of BaYbF$_5$ @SiO$_2$@PEG NPs on cell proliferation was estimated with HeLa and human embryonic kidney 293 (HEK 293) cells (Fig. 5.2d). Under tested concentrations, more than 90% cells remained alive after treatment of BaYbF$_5$@SiO$_2$@PEG NPs. In addition, BaYbF$_5$@SiO$_2$@PEG treatment did not induce the morphological changes (Fig. 5.5).

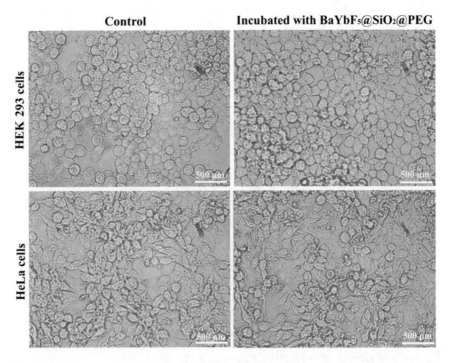

**Fig. 5.5** Microscope images of the HEK 293 and HeLa cells incubated without and with BaYbF$_5$@SiO$_2$@PEG for 24 h. (Copyright 2012 Wiley-VCH)

### 5.3.3 CT Contrast Efficiency

Prior to in vivo imaging, we first compared the contrast efficiency of $BaYbF_5@SiO_2@PEG$ with Iobitridol and $NaYbF_4$–based contrast agent that contains a single contrast element described in previous chapter. The results showed that at 120 KVp, the commonly used operating voltage in the clinic scanning, the CT value varied linearly as a function of the concentration of the contrast element (Fig. 5.6). Notably, at equal concentrations of each agent, the CT value of $BaYbF_5@SiO_2@PEG$ was significantly enhanced compared to Iobitridol and even much higher than that of $NaYbF_4$-based contrast agent. More impressively, at the voltages of 80–140 KVp, $BaYbF_5@SiO_2@PEG$ showed the highest contrast efficiency among these three contrast agents. This is understandable. As we discussed above, the K-edge value of Yb (61 keV) located within the high-energy region of the X-ray spectrum, which dominantly contributes to the X-ray attenuation of $BaYbF_5@SiO_2@PEG$ with the assistance of Ba at higher voltage. When the

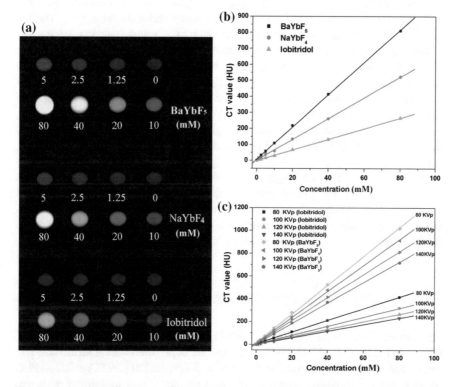

**Fig. 5.6 a** In vitro CT images of $BaYbF_5@SiO_2@PEG$, $NaYbF_4@PEG$ and Iobitridol with different concentrations. **b** CT values (HU) of $BaYbF_5@SiO_2@PEG$, $NaYbF_4@PEG$ and Iobitridol as a function of the agent concentrations at 120 KVp, and **c** at various voltages. (Copyright 2012 Wiley-VCH)

operating voltage is 80 KVp, the high-energy region of the X-ray spectrum will close to the K-edge value of Ba (37 keV), the X-ray attenuation of Ba will be the dominant contribution with the assistant of Yb. The two aspects allow BaYbF$_5$@SiO$_2$@PEG to show much higher contrast efficiency than the other two contrast agents at different operating voltages.

### 5.3.4  In Vivo CT Imaging

Given these promising in vitro results, we proceeded to assess their in vivo CT imaging performance. Figure 5.7a-c are the coronal view and 3D-renderings CT images of the rat after receiving intravenous injection of BaYbF$_5$@SiO$_2$@PEG (800 μL, 0.2 M BaYbF$_5$). Compared to the CT images of NaYbF$_4$@PEG NPs in chapter 4, BaYbF$_5$@SiO$_2$@PEG showed different in vivo imaging performance and a longer circulation time in the blood was found. As seen from Table 5.1, strong CT signals in different vessels were observed upon injection of BaYbF$_5$@SiO$_2$@PEG. A more careful look at Fig. 5.8 revealed clear great vessels including the aortic arch, inferior vena cava, jugular vein, carotid artery and subclavian vein. The CT signal intensity at the heart reached a maximum at 1 h post-injection, which was 5-fold enhanced as compared to preinjection. Even at 2 h postinjection, the bright CT signal remains high, which has never been achieved in previous reports and suggests the great potential of BaYbF$_5$@SiO$_2$@PEG in as a blood pool imaging for the diagnosis of many diseases such as the myocardial infarction, atherosclerotic plaque and thrombosis [7].

To better demonstrate the blood pool imaging of BaYbF$_5$@SiO$_2$@PEG, BaYbF$_5$@SiO$_2$@PEG NPs were injected into the rabbit and the whole body CT imaging was carried out. As shown in Fig. 5.7d-g, various blood vessels were clearly visualized, and the bright signal of these blood vessels remained at 1 h postinjection, even the smaller-sized vessels in the abdomen. Such a long blood circulation time clearly suggests that BaYbF$_5$@SiO$_2$@PEG can effectively accumulate in the tumor tissues through the enhanced permeability and retention (EPR) effect for tumor detection and imaging (Fig. 5.9).

### 5.3.5  Biodistribution of BaYbF$_5$@SiO$_2$@PEG

The tissue distribution of BaYbF$_5$@SiO$_2$@PEG in the rat was examined. FITC-labeled BaYbF$_5$@SiO$_2$@PEG NPs were first prepared with uniform size and shape, and were well dispersed in water (Fig. 5.10). BaYbF$_5$(FITC)@SiO$_2$@PEG solution was intravenously administered into a rat. At 24 post injection, the rat was

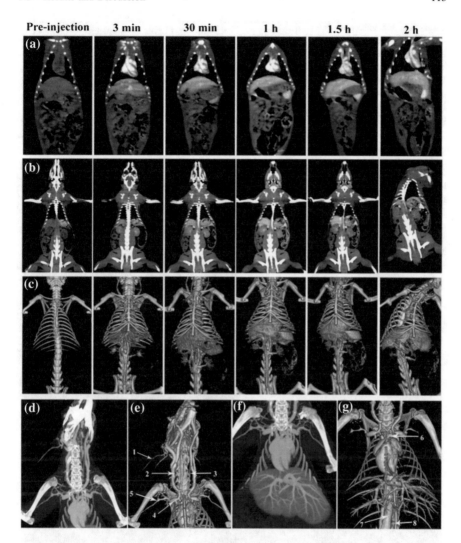

**Fig. 5.7** In vivo CT imaging. (**a,b**) Coronal view images of a rat after intravenous injection of BaYbF$_5$@SiO$_2$@PEG (800 μL, 0.2 M of BaYbF$_5$) solution at timed intervals. **a** Heart and liver. **b** Spleen and kidney. **c** The corresponding 3D-renderings of in vivo CT images. **d–g** In vivo blood pool CT imaging. **d,f** Coronal view images of a rabbit collected at 10 min after intravenous injection of BaYbF$_5$@SiO$_2$@PEG solution. **e,g** The corresponding 3D-renderings of in vivo CT images. The *arrows* indicate several great vessels: 1. auricular vein, 2. jugular vein, 3. Carotid artery, 4. subclavian vein, 5. axillary vein, 6. aortic arch, 7. inferior vena cava, 8. aorta. Besides, atrium and ventricle are also clearly distinguished from each other. (Copyright 2012 Wiley-VCH)

**Table 5.1** CT values of the heart, liver, spleen and kidney of a rat before and after intravenous administration of BaYbF$_5$@SiO$_2$@PEG solution (0.8 mL, 0.2 M) at indicated time intervals. (Copyright 2012 Wiley-VCH)

| Time | Heart | Liver | Spleen | Kidney |
|------|-------|-------|--------|--------|
| Pre-injection | 52.1 | 62 | 52.8 | 41.6 |
| 3 min | 228.9 | 115.7 | 88.3 | 86.1 |
| 30 min | 224.9 | 122.1 | 129.5 | 92.6 |
| 1 h | 260.3 | 142.6 | 178.1 | 91.7 |
| 1.5 h | 202.4 | 146.3 | 178.1 | 81.6 |
| 2 h | 196.6 | 144.4 | 193.6 | 73.5 |

**Fig. 5.8** In vivo CT volume reconstructions image of the rat at 1 h after injection of BaYbF$_5$@SiO$_2$@PEG. The *arrows* indicate several great vessels. (Copyright 2012 Wiley-VCH)

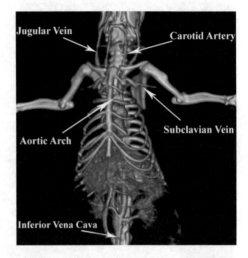

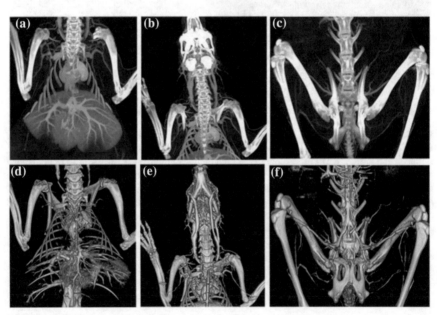

**Fig. 5.9** Blood pool CT imaging. (**a,b,c**) In vivo CT coronal view images of the rabbit collected at 1 h after intravenous injection of BaYbF$_5$@SiO$_2$@PEG solution. (**d,e,f**) The corresponding 3D-renderings of in vivo CT images. (Copyright 2012 Wiley-VCH)

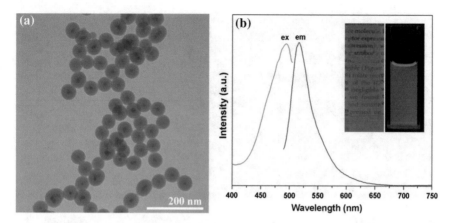

**Fig. 5.10** **a** TEM image of BaYbF$_5$(FITC)@SiO$_2$@PEG. **b** Excitation and emission spectra of BaYbF$_5$(FITC)@SiO$_2$@PEG. The inset shows the corresponding luminescence and brightfield photos of BaYbF$_5$(FITC)@SiO$_2$@PEG in water. (Copyright 2012 Wiley-VCH)

sacrificed and the organs were harvested for fluorescence imaging (Fig. 5.11). Most nanoparticles were cleared from the blood pool and accumulated in liver and spleen through reticuloendothelial system (RES). We also observed the bright fluorescence in the intestines, which was well consistent with CT imaging results as shown in Fig. 5.12 and Table 5.2. These results clearly indicated that BaYbF$_5$@SiO$_2$@PEG can be cleared from the animal body through a hepatobiliary/fecal route [8, 9].

### 5.3.6 In Vivo Side Effects

Finally, the in vivo side effects and elimination of BaYbF$_5$@SiO$_2$@PEG were examined. BaYbF$_5$@SiO$_2$@PEG was intravenously injected into the rat through the tail vein. One month later, the whole body CT imaging was performed to check whether the NPs have been cleared from the body. From Fig. 5.13, we did not see any enhanced CT signals in any organ of the rat, suggesting that BaYbF$_5$@SiO$_2$@PEG has been cleared from the rat body. The organs including the heart, liver, spleen and kidney were also harvested for H&E staining. No tissue damage or any other inflammatory lesion was noticed in the NP-treated rat, demonstrating the low toxicity of BaYbF$_5$@SiO$_2$@PEG. Nevertheless, more in-depth toxicology studies of BaYbF$_5$@SiO$_2$@PEG are still needed.

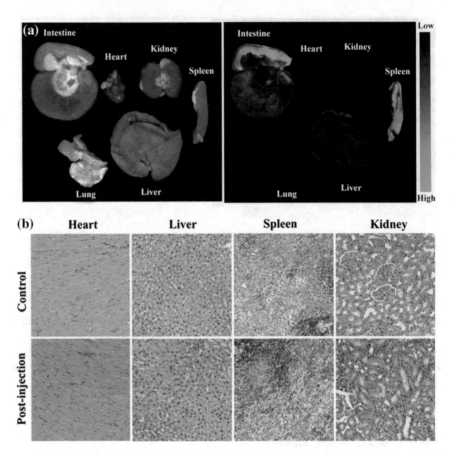

**Fig. 5.11  a** Biodistribution analysis of BaYbF$_5$(FITC)@SiO$_2$@PEG by fluorescence imaging of several organs harvested from the rat 24 after injection of BaYbF$_5$(FITC)@SiO$_2$@PEG solution. *Left*: white light photograph and *right*: fluorescence image of these organs. **b** Histological changes in the heart, liver, spleen and kidney of the rat one month after intravenous injection of a single dose of BaYbF$_5$@SiO$_2$@PEG solution. (Copyright 2012 Wiley-VCH)

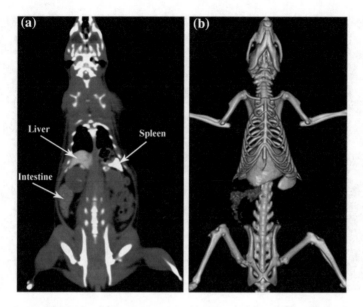

**Fig. 5.12** Biodistribution analysis. (**a,b**) In vivo CT coronal view image and corresponding 3D-renderings of CT image of the rat 24 h after intravenous injection of BaYbF$_5$(FITC) @SiO$_2$@PEG solution. (Copyright 2012 Wiley-VCH)

**Table 5.2** The content of Yb in different organs at 24 h postinjection of FITC-labeled BaYbF$_5$@SiO$_2$@PEG NPs. Notably, whole heart, lung and kidney, whereas parts of the liver, spleen and  intenstine were used for this experiment

| Organs | Heart | Liver | Spleen | Kidney | Lung | Intestine |
|---|---|---|---|---|---|---|
| Yb content | 34.8 μg | 4.8 mg | 4.5 mg | 159 μg | 403 μg | 877 μg |

## 5.4   Conclusion

In this chapter, we have described the first example of binary nanoparticulate CT contrast agent based on BaYbF$_5$@SiO$_2$@PEG nanoparticles. In vitro results demonstrated that BaYbF$_5$@SiO$_2$@PEG significant enhancement of contrast efficacy compared to clinical iodinated agents and the nanoparticulated agents composed of a single contrast element, due to the incorporation of both Ba and Yb with differentiate K-edge energies. After silica coating and PEGylation, the nanoparticles have shown prolonged circulation time in the blood. After intravenous injection, the contrast enhancement in various blood vessels could be clearly observed within 2 h without inducing obvious toxicity in vivo. We expect that BaYbF$_5$@SiO$_2$@PEG may become a promising tool for in vivo X-ray CT angiography.

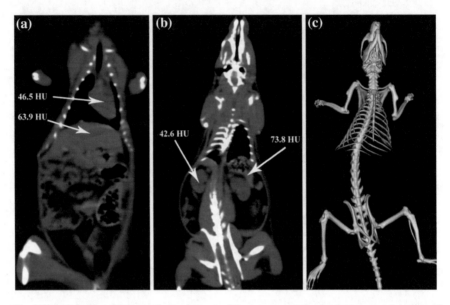

**Fig. 5.13 a,b** In vivo CT coronal view images of the rat one month after intravenous injection of BaYbF$_5$@SiO$_2$@PEG solution. **c** The corresponding 3D-renderings of in vivo CT image. (Copyright 2012 Wiley-VCH)

# References

1. Boyer JC, Vetrone F, Cuccia LA et al (2006) Synthesis of colloidal upconverting NaYF$_4$ nanocrystals doped with Er$^{3+}$, Yb$^{3+}$ and Tm$^{3+}$, Yb$^{3+}$ via thermal decomposition of lanthanide trifluoroacetate precursors. J Am Chem Soc 128:7444–7445
2. Yu CH, Caiulo N, Lo CCH et al (2006) Synthesis and fabrication of a thin film containing silica-encapsulated face-centered tetragonal FePt nanoparticles. Adv Mater 18:2312–2314
3. Yi DK, Selvan ST, Lee SS et al (2005) Silica-coated nanocomposites of magnetic nanoparticles and quantum dots. J Am Chem Soc 127:4990–4991
4. Lu Y, Yin YD, Mayers BT et al (2002) Modifying the surface properties of superparamagnetic iron oxide nanoparticles through a sol—gel approach. Nano Lett 2:183–186
5. Yoon TJ, Kim JS, Kim BG et al (2005) Multifunctional nanoparticles possessing a "magnetic motor effect" for drug or gene delivery. Angew Chem Int Ed 44:1068–1071
6. Piao Y, Burns A, Kim J et al (2008) Designed fabrication of silica-based nanostructured particle systems for nanomedicine applications. Adv Funct Mater 18:3745–3758
7. Kim BH, Lee N, Kim H et al (2011) Large-scale synthesis of uniform and extremely small-sized iron oxide nanoparticles for high-resolution T1 magnetic resonance imaging contrast agents. J Am Chem Soc 133:12624–12631
8. Xiong LQ, Yang TS, Yang Y et al (2010) Long-term in vivo biodistribution imaging and toxicity of polyacrylic acid-coated upconversion nanophosphors. Biomaterials 31:7078–7085
9. Kinsella JM, Jimenez RE, Karmali PP et al (2011) X-ray computed tomography imaging of breast cancer by using targeted peptide-labeled bismuth sulfide nanoparticles. Angew Chem Int Ed 50:12308–12311

# Chapter 6
# Dopamine-Melanin Colloidal Nanospheres for MRI-Guided Photothermal Therapy

**Abstract** Owing to high selectivity and minimal invasiveness, photothermal therapy is emerging as a powerful technique for cancer treatment. However, currently available photothermal therapeutic (PTT) agents have not yet achieved clinical implementation, stemming from great concerns regarding their long-term safety. From this point of view, we develop in this chapter a novel PTT agent based on dopamine-melanin colloidal nanospheres. Benefiting from wide distribution of their component in human naturally, this new PTT agent show biodegradability, a high median lethal dose, and does not induce long-term toxicity during their retention in rats. Moreover, this agent can offer much higher photothermal conversion efficiency than previously reported PTT agents. Upon irradiation with 808 nm laser, dopamine-melanin colloidal nanospheres can efficiently absorb light and transfer it into heat. Both in vitro and in vivo experiments prove that these nanospheres can destroyed tumor tissue and inhibit the regrowth of the tumor. Furthermore, dopamine-melanin colloidal nanospheres can be easily attached to conjugates with other interesting biofunctionalities. By covalent modification of Gd-DTPA, MRI-guided tumor targeted photothermal therapy is achieved.

## 6.1 Introduction

Cancer has overtaken heart diseases and become the leading cause of death worldwide [1]. According to the cancer report, approximately 7.6 million people died of cancer, and 12.7 million new cases of cancer in 2008. 56% of cancer patients were from the developing countries, and the proportion of death of cancer patients from the developing countries has risen to 64% of the total number. In addition, the cancer rates may further increase by 50% to 15 million new cases in the year 2020 according to the World Health Organization. The mortality will continue increasing and it will increase to 70% in the world in 2030 with 30% from the developing countries. Thus, it is highly desirable to develop new therapies and therapeutic agents for more effective treatment of cancer.

© Springer Nature Singapore Pte Ltd. 2018
Y. Liu, *Multifunctional Nanoprobes*, Springer Theses,
DOI 10.1007/978-981-10-6168-4_6

There are two limitations for current clinical cancer treatment. On one hand, the diagnosis and therapy is separate. In this case, it is hard to real-time monitoring the therapeutic effect. On the other hand, current clinical cancer therapies are limited to surgery, radiotherapy, and chemotherapy. Unfortunately, these approaches have many side effects such as killing normal cells, destroying the immune systems, and an increased incidence of second cancers [2, 3]. Thus, the development of novel agents with the abilities to simultaneously detect and treat cancer and minimize the side effects in cancer patients has become increasingly urgent.

Increasing evidence has demonstrated the great potential of photothermal therapy (PTT) as a new and powerful therapy of cancer. Compared to traditional treatment strategies, PTT has many merits including: i) high selectivity. PTT can kill the cells only when PTT agents accumulate in the cells couple with laser irradiation. In another word, no cells will be killed if they have only PTT agent accumulation or laser irradiation alone. ii) This technique also possesses several other potential advantages over traditional techniques, such as facile procedure, faster recovery, fewer complications, and shorter hospital stay [4].

Currently available PTT agents mainly focus on metal nanoparticles (e.g., Au, Ag, and Pd nanoparticles), Cu-based semiconductor nanoparticles, carbon-based nanomaterials and organic polymers. Despite high effectiveness in cancer therapy, these agents have not yet achieved clinical implementation, in a large part due to the safety concerns. For example, metallic nanoparticles have pertinent issues related to the safety of the metal itself, while carbon-based nanomaterials have been demonstrated to trigger many side effects such as oxidative stress and pulmonary inflammation [5, 6]. Given these safety concerns, we hypothesize that an ideal PTT agent should be from the organisms or at least consist of naturally occurring substances in organisms, which will avoid serious adverse effects caused by long-term retention of foreign substances in patients and allow for biodegradation through metabolism. However, it is hard to find such natural materials with intrinsically high absorption in the NIR region and high photothermal conversion efficiency. To the best of our knowledge, only one type of such material has been reported by Zheng et al. as an effective in vivo PTT agent based on porphyrin-lipid [7]. Thanks to their organic nature, the resultant porphysomes were enzymatically biodegradable and induced minimal acute toxicity during their retention in mice. Nevertheless, this agent requires relatively complicated organic synthesis. Thus, it is still challenging to develop such material-based PTT agents with simple preparation processes and high photothermal conversion efficiency.

Melanin is a well-known biopolymer that is widely distributed in the furs, skin, and organs in almost all living organisms including human. It can protect human skin from ultraviolet injury due to its high UV absorption and scavenging of free radicals generated by UV irradiation. Melanin has also thermoregulation function. In addition, its absorption can extend from UV to NIR regions, which inspired us to investigate whether melanin can be an effective PTT agent [8, 9]. In this chapter, we have developed a facile strategy to prepare dopamine-melanin colloidal nanospheres (Dpa-melanin CNSs), followed by surface modification of Gd-chelate for in vivo MRI-guided photothermal therapy of cancer. We have systemically investigated their in vitro and in vivo anticancer effects. The results have demonstrated that Dpa-melanin CNSs could effectively accumulate in tumor tissues and suppress the tumor growth upon irradiation with a NIR laser.

## 6.2 Experimental Section

### 6.2.1 Materials

Dopamine HCl, HAuCl$_4$, cetyltrimethylammonium bromide (CTAB), NaBH$_4$, and AgNO$_3$ were purchased from Alfa. Poly(ethylene glycol) bis(3-aminopropyl) terminated and Diethylenetriaminepentacetic acid bis-anhydride were purchased from Sigma Aldrich.

### 6.2.2 Synthesis of Dpa-Melanin CNSs

Dpa-melanin CNSs were prepared with a modified microemussion method. For preparing Dpa-melanin CNSs with an average diameter of 160 nm, ammonia aqueous solution (2 mL, NH$_4$OH, 28–30%) was mixed with ethanol (40 mL) and deionized water (90 mL) under mild stirring at 30 °C for 30 min. Dopamine hydrochloride (0.5 g) dissolved in deionized water (10 mL) was then injected into the above mixture solution. The color of this solution immediately turned to pale yellow and gradually changed to dark brown. The reaction was allowed to proceed for 24 h. Dpamelanin CNSs were obtained by centrifugation and washed with water for three times. Dpa-melanin CNSs with an average diameter of 70 nm were prepared by increasing the volume of ammonia aqueous solution to 3 mL.

### 6.2.3 Synthesis of Gd-DTPA-Modified Dpa-Melanin CNSs

200 mg as-prepared Dpa-melanin CNSs was dispersed in Tris buffer (10 mM, pH = 8.5). To this solution, 500 mg poly(ethylene glycol) bis(3-aminopropyl) terminated was added. After vigorous stirring for 12 h, the resultant NH$_2$-modified Dpa-melanin CNSs were retrieved by centrifugation, washed with deionized water for three times, and finally dispersed in deionized water. Excessive Diethylenetriaminepentacetic acid bis-anhydride was added in the aqueous solution of NH$_2$-modified Dpa-melanin CNSs and the pH value of this mixture was adjusted to about 7 by adding NaOH. After stirring for 24 h, the DTPA-modified Dpa-melanin CNSs were isolated by centrifugation, washed with deionized water twice, and then redispersed in deionized water. Gd$^{3+}$ complexation is carried out through the

addition of GdCl$_3$ to the DTPA-modified Dpamelanin CNSs colloidal solution under stirring at room temperature for 3 h. The pH value of this solution was adjusted to about 6.5 by adding NaOH.

### 6.2.4 Synthesis of Au Nanorods

Au nanorods were prepared based on the silver ion-assisted seed mediated method according to previous literature [301]. Briefly, the seed solution was synthesized by the addition of HAuCl$_4$ (0.01 M, 0.25 mL) into CTAB (0.1 M, 10 mL). After gentle mixing, a freshly prepared, ice-cold NaBH$_4$ solution (0.01 M, 0.6 mL) was then added into the mixture solution under vigorous stirring for 2 min. The seed solution was kept at room temperature for 2 h without disturbance. To grow Au nanorods, HAuCl$_4$ (0.01 M, 4.0 mL) and AgNO$_3$ (0.01 M, 0.8 mL) were injected into CTAB solution (0.1 M, 80 mL), followed by the addition of HCl (1.0 M, 1.6 mL). After gentle mixing of the solution, ascorbic acid (0.1 M, 0.64 mL) was then added. Finally, the seed solution (192 μL) was injected into the growth solution, and the solution was gently mixed for 30 s and left undisturbed at 27–30 °C for 6 h. After removal of excess CTAB by centrifugation and water washing, PEG-stabilized Au nanorods were prepared by ligand exchange.

### 6.2.5 Cytotoxicity Assay of Dpa-Melanin CNSs

The viability and proliferation of 4T1 cells were evaluated by a wellknown methyl thiazolyl tetrazolium (MTT) assay. Typically, 4T1 cells were incubated in the culture medium at 37 °C in an atmosphere of 5% CO$_2$ and 95% air for 24 h. Subsequently, the culture medium was replaced with fresh culture medium containing different concentrations of Dpa-melanin CNSs for another 24 h. The cells were then washed with medium twice. 100 μL of the new culture medium containing MTT reagent (10%) was added to each well of the 96-well assay plate and incubated for 4 h to allow formation of formazan dye. After removal of the medium, the purple formazan product was dissolved with DMSO for 15 min. Finally, the optical absorption of formazan at 570 nm was measured by an enzyme-linked immunosorbent assay reader.

### 6.2.6 Measurement of Photothermal Performance

1 mL of aqueous dispersion of Dpa-melanin CNSs with different concentrations (0–200 μg/mL) were introduced in a quartz cuvette and irradiated with an 808 nm NIR laser at a power density of 2 W/cm$^2$ for 500 s, respectively. A thermocouple probe

with an accuracy of 0.1 °C was inserted into the Dpa-melanin CNSs aqueous solution perpendicular to the path of the laser. The temperature was recorded every 10 s by a digital thermometer with a thermocouple probe.

The phothermal conversion efficiency of Au nanorods was determined at the same conditions as those for Dpa-melanin CNSs.

## 6.2.7  In Vitro Photothermal Cytotoxicity

Photothermal cytotoxicity of Dpa-melanin CNSs was examined on 4T1 murine breast cancer cells and HeLa cells. Briefly, 4T1 and HeLa cells were incubated in 6-well plates at 37 °C containing Dpa-melanin CNSs at a concentration of 200 µg/mL for 30 min, and then irradiated with an 808 nm laser at a power density of 2 W/cm$^2$ for 5 min. The cells were stained with both calcein AM (calcein acetoxymethyl ester) and PI (propidium iodide).

To quantify the photothermal cytotoxicity of Dpa-melanin CNSs, 4T1 cells were incubated in 96-well plates at 37 °C in a humidified atmosphere containing 5% $CO_2$ for 24 h. Dpa-melanin CNSs with different concentrations were added and the cells were incubated for 30 min. Thereafter, the cells were exposed to an NIR laser (808 nm, 2 W/cm$^2$) for 5 min, and then incubated for another 24 h. The viability of 4T1 cells was evaluated by an MTT assay.

## 6.2.8  Tumor Model

Animal care and handing procedures were in agreement with the guidelines of the Regional Ethics Committee for Animal Experiments.

4T1 cells were cultured in RPMI-1640 supplemented with 10% FBS at 37 °C in an atmosphere of 5% $CO_2$ and 95% air. $2 \times 10^6$ 4T1 cells suspended in 100 µL serum free cell medium were inoculated subcutaneously in several female Balb/c mice. The mice were used for further experiments when the tumor had grown to 3–4 mm in diameter.

## 6.2.9  In Vivo Photothermal Cytotoxicity

For observation of in vivo photothermal therapy effect of Dpa-melanin CNSs, 4T1 tumor-bearing mice were first anesthetized by intraperitoneal injection of chloral hydrate solution (10 wt%) and intratumorally injected with 100 µL Dpa-melanin CNSs aqueous dispersion (200 µg/mL). Subsequently, the tumors were irradiated with an 808 nm laser at 2 W/cm$^2$ for 5 min. The tumor sizes were measured by a caliper every the other day for 10 days and calculated as the volume.

$$\text{Tumor volume} = (\text{tumor length}) \times (\text{tumor width})^2/2. \qquad (6.1)$$

Relative tumor volumes were calculated as $V/V_0$ ($V_0$ was the tumor volume when the treatment was initiated). To examine the histological changes of the tumors, some tumor-bearing mice were killed after the laser irradiation, and the tumors were removed and stained with Hematoxylin and eosin (H&E) for histopathology analysis.

### 6.2.10  Blood Analysis

Healthy rats were intravenously injected with a single dose of Dpamelanin CNSs. Several other rats were used as the controls. Over one month period, the rats were observed for behavioral changes and the weight was also monitored. At the one-month point, the rats were anesthetized and the blood was collected by a cardiac puncture method for blood biochemistry assay. Several organs including the heart, liver, lung, kidney, and spleen were preserved in a 10% formalin solution and used for histological analysis.

### 6.2.11  In Vivo MR Imaging

For determination of the specific relaxivity values ($r1$), Gd-DTPA modified Dpa-melanin CNSs were dispersed in water with Gd concentrations in the range of 0–0.4 mM. For in vivo MRI, the 4T1 tumor bearing mouse was first anesthetized by intraperitoneal injection of chloral hydrate solution (5 wt%), and then intravenously injected with 200 $\mu$L of Gd-DTPA-modified Dpa-melanin CNSs aqueous solution at the Gd concentration of 12 mM. The $T1$-weighted images were acquired using a 1.5 T human clinical scanner.

### 6.2.12  Biodistribution of Dpa-Melanin CNSs

Gd-DTPA-modified Dpa-melanin CNSs solution was injected intravenously into 4T1 tumor-bearing mice. After 24 h, the mice were sacrificed and the concentration of Gd in several organs including heart, liver, spleen, kidney, lung, intestine, and tumor was monitored by ICP-MS analysis. For ICP-MS analysis, these organs were lyophilized, weighed, and digested in aqua regia for 2 h at 80 °C to dissolve the tissues. The concentrations of Gd in these organs were quantified by ICP-MS.

### 6.2.13 Resonance Light Scattering of Dpa-Melanin CNSs and Au Nanorods

Dpa-melanin CNSs and Au nanorods were dispersed in water, and were adjusted to the same absorbance at 808 nm. The resonance light scattering spectra of Dpa-melanin CNSs and Au nanorods were recorded on a Perkin–Elmer LS 55 luminescence spectrometer. Excitation and emission were set to the same wavelength and scanned from 600 nm to 780 nm.

### 6.2.14 Characterization

The morphology of Dpa-melanin CNSs was characterized by using a TECNAI G2 high-resolution transmission electron microscope and a FEI/Philips XL30 ESEM FEG field-emission scanning electron microscope. XPS measurements were conducted with a VG ESCALAB MKII spectrometer. The XPSPEAK software (Version 4.1) was used to deconvolute the narrow-scan XPS spectra of the C 1 s, N 1 s and O 1 s of the samples, using adventitious carbon to calibrate the C 1 s binding energy (284.5 eV). 13C NMR analyses were carried out on a Varian Infinityplus 400 spectrometer operating at a magnetic field strength of 9.4 T. FTIR analysis was performed on a Bruker Vertex 70 spectrometer ($2 \ cm^{-1}$). Raman spectra were recorded on a J–Y T64000 Raman spectrometer with 514.5 nm wavelength incident laser light.

## 6.3 Results and Discussion

### 6.3.1 Synthesis and Characterization of Dpa-Melanin CNSs

Dpa-melanin CNSs were synthesized by the oxidation and self-polymerization of dopamine in a mixture containing water, ethanol, and ammonia at room temperature. Compared to previously reported polymer-based PTT agents, our strategy is straightforward under mild conditions and without the need of large and complicate instruments. SEM and TEM images showed that Dpa-melanin CNSs were uniform in shape with an average diameter of 160 nm (Figs. 6.1a, b and 6.2a, b). The size of Dpa-melanin CNSs could be well controlled by tuning the molar ratio between ammonia to dopamine. Smaller Dpa-melanin CNSs with an average diameter of 70 nm were obtained when the ratio was increased to 17. The small Dpa-melanin CNSs remained well dispersed in water as shown in Fig. 6.3. However, we found prolonging the reaction time did not significantly affect the diameter of Dpa-melanin CNSs (Fig. 6.4) with a slight increase. Indeed, our Dpa-melanin

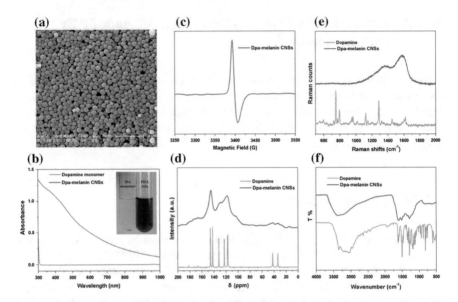

**Fig. 6.1 a** SEM image of Dpa-melanin CNSs. **b** UV–vis absorption spectra of dopamine and Dpa-melanin CNSs. The inset shows a photograph of dopamine and Dpa-melanin CNS aqueous solutions. **c** ESR spectrum of Dpa-melanin CNSs. **d** $^{13}$C NMR spectra. **e** Raman spectra. **f** FTIR spectra of dopamine and Dpa-melanin CNSs, respectively. (Copyright 2013 Wiley-VCH)

CNSs are closed to the naturally occurring melanin regarding the size and shape, which typically shows 40–150 nm in diameter and spherical in shape [10].

One distinguish feature of melanin from other polymers is the paramagnetism because of the presence of stable $\pi$-electron free radicals [11, 12]. We performed electron spin resonance (ESR) analysis to confirm the successful synthesis of melanin. Similar to the naturally occurring melanin [13, 14], the synthesized Dpa-melanin CNSs showed a single-line ESR spectrum, and a single peak was observed with a g-factor approaching two (Fig. 6.1c). Moreover, solid-state $^{13}$C NMR, Raman, Fourier transform infrared and X-ray photoelectron spectra of Dpa-melanin CNSs were consistent with those of naturally occurring melanin. In addition, the as-prepared Dpa-melanin CNSs have high stability and remain stable even after one month without obvious aggregation. Even after dispersal in 10% blood serum solution, no change in their absorption was observed after 24 h (Figs. 6.5 and 6.6b). All these results clearly indicated their great potential for in vivo applications.

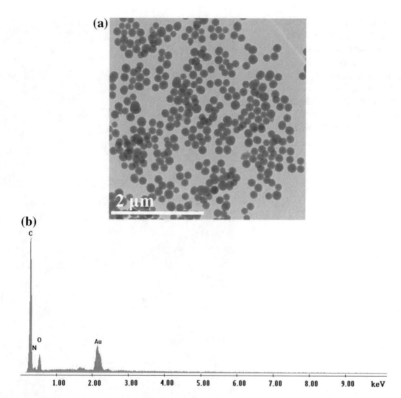

**Fig. 6.2 a** TEM image of Dpa-melanin CNSs with an average diameter of 160 nm. **b** Energy-dispersive X-ray analysis (EDAX) spectra of Dpa-melanin CNSs. (Copyright 2013 Wiley-VCH)

## 6.3.2   Photothermal Conversion Efficacy

Photothermal therapy employs photosensitizing agents to generate heat from light absorption at the target sites. To avoid damage to the healthy cells, photosensitizing agents must have high absorption in the NIR region of the light spectrum, owing to the deep penetration of NIR and its low absorption by human tissues. Impressively, our Dpa-melanin CNSs exhibited broad absorption ranging from ultraviolet (UV) to NIR wavelengths relative to dopamine monomer, and the absorbance increases linearly with the concentration of Dpa-melanin CNSs (Figs. 6.1b and 6.7). The absorption in the UV region was attributed to the oxidation of dopamine into dopachrome and dopaindole, and the following self-polymerization process led to an absorption extension to NIR wavelengths. We expect this strong absorption in the NIR region allows Dpa-melanin CNSs to act as an effective PTT agent.

In order to quantitatively assess the photothermal conversion efficiency, we calculated the molar extinction coefficient $\varepsilon_{808}$ according to the Eq. 6.2 [15]:

**Fig. 6.3  a, b** SEM and TEM images of Dpa-melanin CNSs with an average diameter of about 70 nm, respectively. **c** Hydrodynamic size of Dpa-melanin CNSs determined by dynamic light scattering (DLS) analysis. (Copyright 2013 Wiley-VCH)

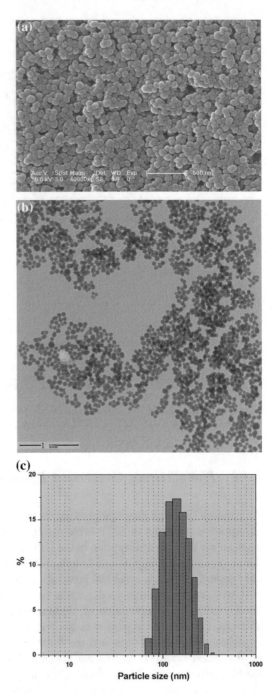

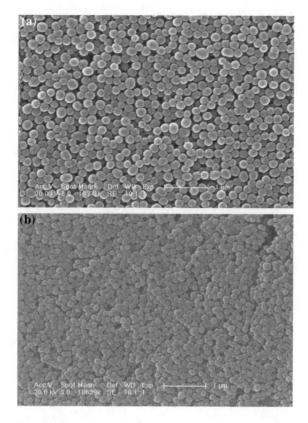

**Fig. 6.4** **a** SEM image of Dpa-melanin CNSs obtained with 36 h of reaction time and 2 mL of ammonia. **b** SEM image of Dpa-melanin CNSs obtained with 36 h of reaction time and 3 mL of ammonia. (Copyright 2013 Wiley-VCH)

$$\varepsilon = (A\ V_{NPs}\ \rho\ N_A)/(LC) \tag{6.2}$$

Where $\varepsilon$ is the extinction coefficient, A is the absorption of nanoparticles at a wavelength of 808 nm, $V_{NPs}$ (in cm$^3$) is the average volume of the nanoparticles, $\rho$ is the density of nanoparticles, $N_A$ is Avogadro's constant, L is the path length (1 cm), and C (in g/L) is the weight concentration of the nanoparticles.

According to the equation, the $\varepsilon_{808}$ was determined to $7.3 \times 10^8$ M$^{-1}$ cm$^{-1}$. Although this value was a little lower than that of porphysomes ($\approx 10$ 9 M$^{-1}$ cm$^{-1}$), it is still larger than those of many other PTT agents, as summarized in Table 6.1.

Another crucial parameter is the photothermal conversion efficiency. We next evaluated the photothermal conversion effects of different concentrations of Dpa-melanin CNSs and calculated the photothermal conversion efficiency. Figure 6.8a shows the temperature increasing curves of different concentrations of Dpa-melanin CNSs. It is clear that the temperature of Dpa-melanin CNS solution increased with the irradiation time, and it increased by 33.6 °C after 500 s laser irradiation at a concentration of 200 µg/mL with obvious water drop on the wall of the quartz vessel. In contrast, only 3.2 °C increase was found for the pure water after laser irradiation. It has been suggested that the cancer cells can be killed after

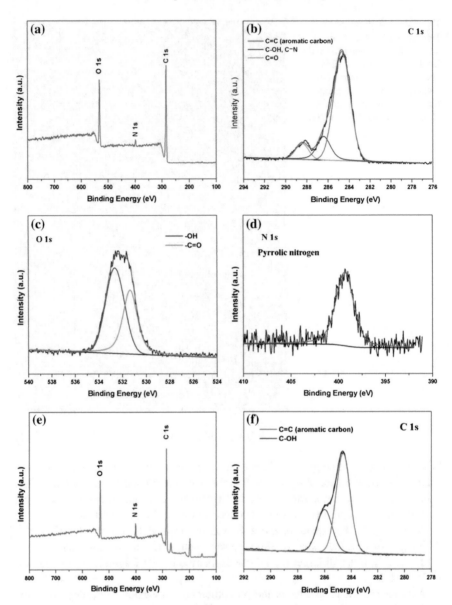

**Fig. 6.5** XPS analysis of Dpa-melanin CNSs and dopamine monomer. **a–d** XPS survey spectrum, C 1s, O 1s and N 1s XPS spectra of Dpa-melanin CNSs. **e–g** XPS survey spectrum, C 1s and N 1s XPS spectra of dopamine monomer. (Copyright 2013 Wiley-VCH)

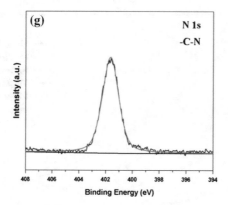

**Fig. 6.5** (continued)

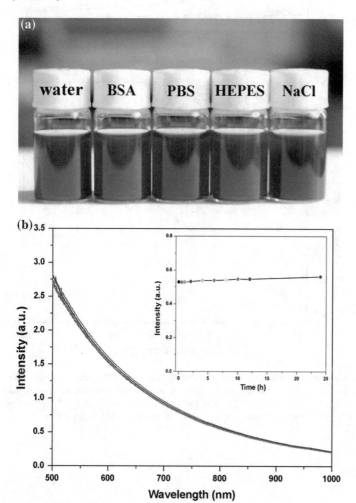

**Fig. 6.6** **a** Dispersibility of Dpa-melanin CNSs in water and different buffers. **b** Colloidal stability of Dpa-melanin CNSs dispersion in 10% of serum solution. The inset shows the absorbance at 808 nm wavelength versus time. (Copyright 2013 Wiley-VCH)

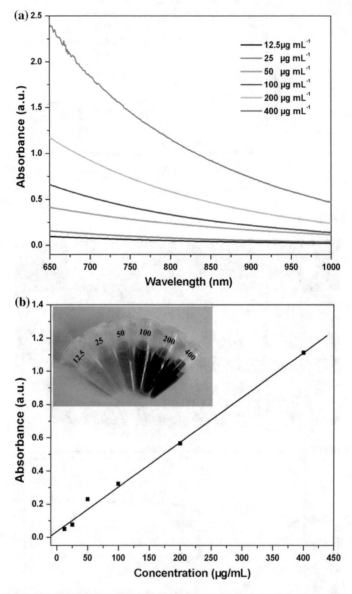

**Fig. 6.7 a** Room temperature UV-vis absorbance spectra of Dpa-melanin CNSs with different concentrations. **b** A linear relationship for the optical absorbance at 808 nm as a function of the concentration of Dpa-melanin CNSs. The inset shows the photograph of Dpa-melanin CNSs dispersion in water with different concentrations. (Copyright 2013 Wiley-VCH)

**Table 6.1**  Molar extinction coefficients of several PTT agents

| References | PTT agents | Molar Extinction Coefficient ($M^{-1}$ $cm^{-1}$) | Wavelength (nm) |
|---|---|---|---|
| This study | Dpa-melanin CNSs | $7.3 \times 10^8$ | 808 |
| *Photochem. Photobiol.* **1998**, *68*, 141–142 | Rhodamine 6G | $1.2 \times 10^5$ | 530 |
| *Photochem. Photobiol.* **1998**, *68*, 141–142 | Malachite Green | $1.5 \times 10^5$ | 617 |
| *Chem. Mater.* **2003**, *15*, 2854–2860 | CdX (X = S, Se, Te) | $2.5 \times 10^5$ | At excitonic maximum |
| *Proc. Natl. Acad. Sci. U.S.A.* **2005**, *102*, 11600–11605 | Carbon nanotubes | $7.9 \times 10^6$ | 808 |
| *Nano Lett.* **2011**, *11*, 2560–2566 | Copper selenide | $7.7 \times 10^7$ | 970 |
| *Nature Mater.* **2011**, *10*, 324–332 | Porphysome | $2.9 \times 10^9$ | 680 |

maintenance at 42 °C for 15–60 min; this duration can be shortened to 4–6 min for temperatures over 50 °C [16]. Assuming that the temperature of human body is 36 °C, after injection of Dpa-melanin CNSs, the tumor tissues can easily be heated to over 50 °C within 5 min after laser irradiation, thus efficiently killing the cancer cells.

According to the following equations, we calculated the photothermal conversion efficiency of Dpa-melanin CNSs:

Based on the total energy balance for this system [17]:

$$\sum_i m_i \, c_{p,i} \, \frac{dT}{dt} = Q_{NPs} + Q_s - Q_{loss} \tag{6.3}$$

Where $m$ and $C_p$ are the mass and heat capacity of solvent (water), respectively. T is the solution temperature.

$Q_{NPs}$ is the photothermal energy input by Dpa-melanin CNSs:

$$Q_{NPs} = I(1 - 10^{-A_\lambda})\eta \tag{6.4}$$

where $I$ is the laser power, $A_\lambda$ is the absorbance of Dpa-melanin CNSs at the wavelength of 808 nm, and $\eta$ is the conversion efficiency from the absorbed light energy to thermal energy. $Q_s$ is the heat associated with the light absorbance of the solvent, which is measured independently to be 25.2 mW using pure water without Dpa-melanin CNSs.

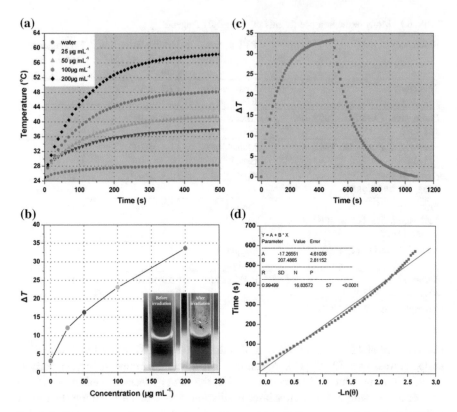

**Fig. 6.8  a** Temperature elevation of water and Dpa-melanin CNS aqueous solutions with different concentrations as a function of irradiation time. **b** Plot of temperature change ($\Delta T$) over a period of 500 s versus the concentration of Dpa-melanin CNSs. The inset shows the photograph of Dpa-melanin CNSs dispersion in water before and after laser irradiation. **c** The photothermal response of the Dpa-melanin CNSs aqueous solution (200 µg/mL) for 500 s with an NIR laser (808 nm, 2 W/cm²) and then the laser was shut off. **d** Linear time data versus—ln θ obtained from the cooling period of c. (Copyright 2013 Wiley-VCH)

$Q_{loss}$ is thermal energy lost to the surroundings:

$$Q_{loss} = hA\Delta T \tag{6.5}$$

where h is the heat transfer coefficient, A is the surface area of the container, and $\Delta T$ is the temperature change, which is defined as T-$T_{surr}$ (T and Tsurr are the solution temperature and ambient temperature of the surroundings, respectively).

At the maximum steady-state temperature, the heat input is equal to the heat output, that is:

$$Q_{NPs} + Q_s = Q_{loss} = hA\Delta T_{max} \tag{6.6}$$

where $\Delta T_{max}$ is the temperature change at the maximum steady-state temperature. According to the Eqs. 6.2 and 6.4, the photothermal conversion efficiency ($\eta$) can be determined:

$$\eta = \frac{hA\Delta T_{max} - Q_s}{I(1 - 10^{-A_\lambda})} \tag{6.7}$$

In this equation, only $hA$ is unknown for calculation. In order to get the $hA$, we herein introduce $\theta$, which is defined as the ratio of $\Delta T$ to $\Delta T_{max}$:

$$\theta = \frac{\Delta T}{\Delta T_{max}} \tag{6.8}$$

Substituting Eq. 6.8 into Eq. 6.3 and rearranging Eq. 6.8:

$$\frac{d\theta}{dt} = \frac{hA}{\sum_i m_i \, C_{p,i}} \left[ \frac{Q_{NPs} + Q_s}{hA\Delta T_{max}} - \theta \right]. \tag{6.9}$$

When the laser was shut off, the $Q_{NPs} + Q_s = 0$, Eq. 6.9 changed to Eq. 6.10:

$$dt = -\frac{\sum_i m_i \, C_{p,i}}{hA} \frac{d\theta}{\theta} \tag{6.10}$$

Integrating Eq. 6.10 gives the expression Eq. 6.11:

$$t = -\frac{\sum_i m_i \, C_{p,i}}{hA} \theta \tag{6.11}$$

Thus, $hA$ can be determined by applying the linear time data from the cooling period vs $-\ln\theta$ (Figs. 6.9 and 6.10). Substituting $hA$ value into Eq. 6.7, the photothermal conversion efficiency ($\eta$) of Dpa-melanin CNSs was calculated to be 40%. In contrast, the $\eta$ of Au nanorods is only 22%.

In order to reveal the possible reasons for such a high photothermal conversion capability of Dpa-melanin CNSs, we investigated their absorption and photostability, two critical factors in determining the $\eta$ values of PTT agents. Figures 6.11 and 6.12 are the photostability of Dpa-melanin CNSs and Au nanorods. Clearly, Dpa-melanin CNSs have much higher photostability than Au nanorods, and no change in the absorbance was found even after continuous laser irradiation for 1 h. SEM images showed no changes in size and morphology (Fig. 6.11c). On the contrary, Au nanorods suffered significant loss of the NIR absorbance after laser irradiation and morphological change (melting) after laser irradiation.

The high photothermal conversion efficiency of Dpa-melanin CNSs prompted us to evaluate their feasibility as a PTT agent for cancer therapy. 4T1 and HeLa cells were incubated with Dpa-melanin CNSs for 30 min and then exposed to an 808 nm

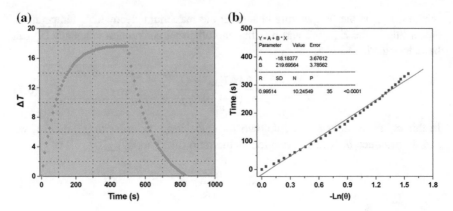

**Fig. 6.9** **a** The photothermal response of the aqueous dispersion of Au nanorods for 500 s with an NIR laser (808 nm, 2 W/cm²) and then the laser was shut off. **b** Linear time data versus -lnθ obtained from the cooling period of. (Copyright 2013 Wiley-VCH)

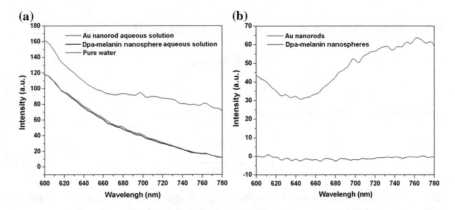

**Fig. 6.10** **a** Resonance light scattering spectra of gold nanorods aqeous solution, Dpa-melanin CNSs aqueous solution and pure water. **b** Resonance light scattering spectra of gold nanorods and Dpa-melanin CNSs after blank subtraction. (Copyright 2013 Wiley-VCH)

laser at 2 W/cm² for 5 min, followed by staining with PI and calcein AM (Fig. 6.13a–d). Under excitation, a clear demarcation line between the regions of live cells (green) and dead cells (red) was observed, and almost all the cells within the laser spot were killed. In contrast, cells treated with either Dpa-melanin CNSs or laser alone shown negligible cell death (Fig. 6.14). Similar results were obtained for HeLa cells (Fig. 6.15 and 6.16), demonstrating that Dpa-melanin CNSs hold great promise as an effective PTT agent for treatment of different cancers.

MTT assay was then carried out to quantify the phothermal effect. 4T1 cells were incubated with increasing concentrations of Dpa-melanin CNSs for 24 h, and we did not find obvious cell death for cells treated with Dpa-melanin CNSs. Even at

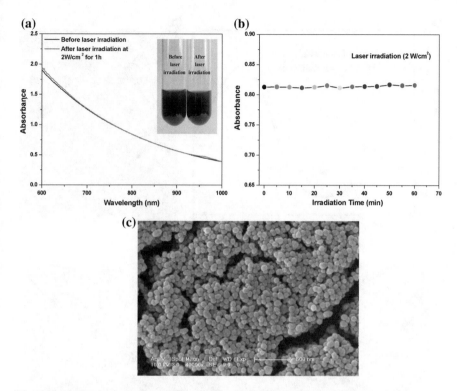

**Fig. 6.11** **a** UV-vis absorption spectra of Dpa-melanin CNSs dispersion in water before versus after laser irradiation for 1 h (808 nm, 2 W/cm$^2$). The inset shows the photograph of Dpa-melanin CNSs dispersion in water before versus after laser irradiation for 1 h. **b** Optical absorbance at 808 nm versus the irradiation time. **c** SEM image of Dpa-melanin CNSs after laser irradiation for 1 h. (Copyright 2013 Wiley-VCH)

the highest tested dose of Dpa-melanin CNSs (1.2 mg/mL), the cell viability still remained approximately 90% (Fig. 6.13e). However, upon laser irradiation, less than 20% of cells remained alive at a concentration of 200 μg/mL.

## 6.3.3 In Vivo Photothermal Therapy

Next, the in vivo photothermal therapy effect of Dpa-melanin CNSs was investigated. 4T1 tumor-bearing mice were intratumorally injected with the aqueous solution of Dpa-melanin CNSs and the tumors were exposed to an 808 nm laser for 5 min at a power density of 2 W/cm$^2$. Most of the tumor tissue was necrotic after treatment; shrunken malignant cells, cytoplasmic acidophilia, and corruption of the tumor extracellular matrix were observed in the H&E staining pictures (Fig. 6.17). In contrast, no tissue damage was found for mice treated with Dpa-melanin CNSs

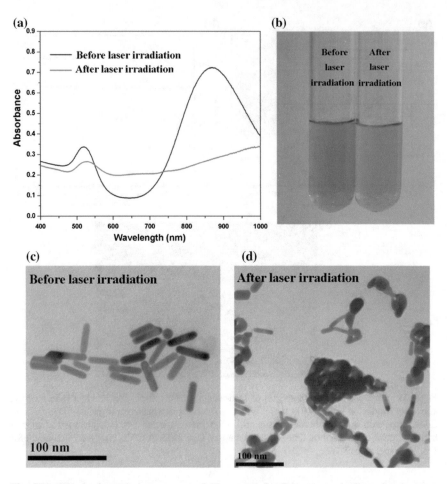

**Fig. 6.12** UV-vis absorption spectra **a** and Photograph **b** of Au nanorods dispersion in water before and after laser irradiation. TEM images of Au nanorods before **c** and after **d** laser irradiation. (Copyright 2013 Wiley-VCH)

or laser alone. We also monitored the tumor growth within 10 days, and the tumors in mice treated with Dpa-melanin CNSs were ablated after photothermal treatment without regrowth or with rather slow growth. In contrast, tumors in the control animals and mice treated with laser alone continued to grow rapidly (Fig. 6.18).

### 6.3.4   MRI Imaging

Dpa-melanin CNSs are easy to be modified with thiol- and amino-terminated molecules through the Michael addition or Schiff base reaction, thus providing an

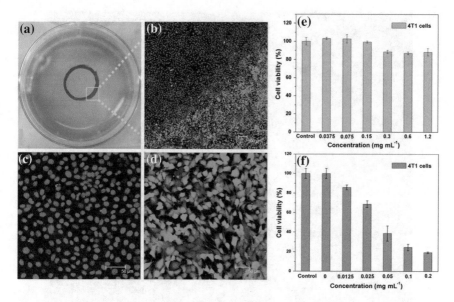

**Fig. 6.13  a** A digital photo of the 4T1 cells-containing culture dish after incubation with Dpa-melanin CNSs. The red circle shows the laser spot. **b–d** Confocal images of calcein AM (*green*, live cells) and propidium iodide (*red*, dead cells) co-stained 4T1 cells after laser irradiation. Cell viability of 4T1 cells after incubation with increased concentrations of Dpa-melanin CNSs without laser irradiation **e** versus after laser irradiation (808 nm, 2 W/cm$^2$, 5 min). (Copyright 2013 Wiley-VCH)

important platform for in vivo applications. For a proof-of-concept study, we fabricated Gd-DTPA modified Dpa-melanin CNSs and further investigated their applications in MRI and therapy of tumor in vivo. Figure 6.19a showed characteristic peaks corresponding to PEG, confirming the successful conjugation of PEG on the surface of Dpa-melanin CNSs. The Gd-DTPA modified Dpa-melanin CNSs possessed a relaxivity value of 6.9 mM$^{-1}$ s$^{-1}$, superior to the commercial Magnevist (Fig. 6.19b). After intravenous injection, there was a time-dependent T1 signal increase at the tumor site and the signal became quite strong at 24 h postinjection (Fig. 6.20). The accumulation of Gd-DTPA modified Dpa-melanin CNSs in tumor was 5.7% ID/g (Fig. 6.21). Upon laser irradiation, the tumor was also destroyed (Figs. 6.22 and 6.23).

## 6.3.5   In Vivo Toxicity Studies of Dpa-Melanin CNSs

Melanin is widely distributed in human body. Thus, we hypothesized that Dpa-melanin CNSs should have good biocompatibility and biodegradation. In our preliminary observations, we found that Dpa-melanin CNSs lost their absorbance along with color fading with H$_2$O$_2$ (Fig. 6.24). H$_2$O$_2$ is an endogenous molecule

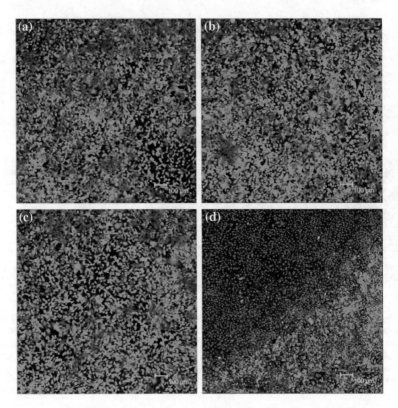

**Fig. 6.14** Confocal microscopic images of differently treated 4T1 cells stained with calcein AM (*green*, live cells) and propidium iodide (*red*, dead cells): **a** control; **b** laser irradiation only; **c** Dpa-melanin CNSs only; and **d** with the treatment of Dpa-melanin CNSs along with laser irradiation. (Copyright 2013 Wiley-VCH)

produced by reduced nicotinamide adenine dinucleotide phosphate (NADPH) oxidases in phagocytes and many organs [18]. To confirm the low toxicity of Dpa-melanin CNSs, the median lethal dose (LD 50), a standardized measure used to evaluate the acute toxicity of agents, was examined. The result showed that the LD50 of Dpa-melanin CNSs is 483.95 mg/kg with a 95% confidence interval of 400.22 to 585.19 mg/kg (intravenous injection). This dose was nearly five hundred times higher than the dose used for photothermal therapy in this work. Next, Dpa-melanin CNSs were intravenously injected into rats, and body weight was monitored over one-month period. As shown in Fig. 6.25, the body weight of the treated group gradually increased in a manner similar to that of the control group. In addition, no abnormalities were observed in eating, drinking, grooming, activity, exploratory behavior, urination, or neurological status (Fig. 6.26).

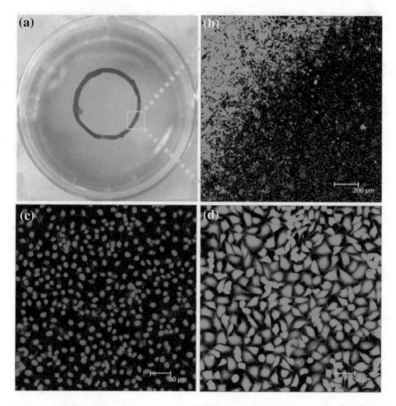

**Fig. 6.15 a** A digital photo of the HeLa cell culture dish after incubation with Dpa-melanin CNSs. The red circle shows the laser spot. **b–d** Confocal images of calcein AM (*green*, live cells) and propidium iodide (*red*, dead cells) co-stained cells after laser irradiation. (Copyright 2013 Wiley-VCH)

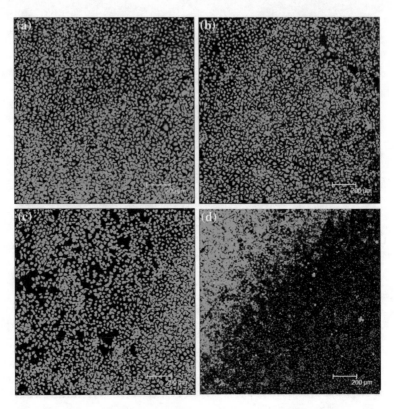

**Fig. 6.16** Confocal microscopic images of differently treated HeLa cells stained with calcein AM (*green*, live cells) and propidium iodide (*red*, dead cells): **a** control; **b** laser irradiation only; **c** Dpa-melanin CNSs only; and **d** with the treatment of Dpa-melanin CNSs along with laser irradiation. (Copyright 2013 Wiley-VCH)

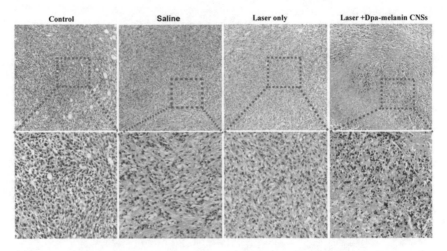

**Fig. 6.17** Representative Hematoxylin and eosin (H&E) stained histological images of tumor sections at different magnifications from control group and the mice that received different treatments. (Copyright 2013 Wiley-VCH)

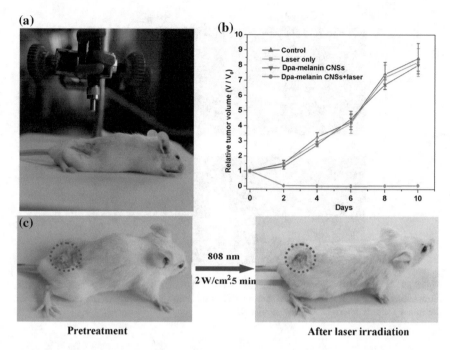

**Fig. 6.18** **a** Photothermal therapy set-up showing laser and the 4T1 tumor-bearing mouse. **b** Time-dependent tumor growth curves of the mice after different treatments. **c** Digital photos of a 4T1 tumor-bearing mouse before and after photothermal therapy. (Copyright 2013 Wiley-VCH)

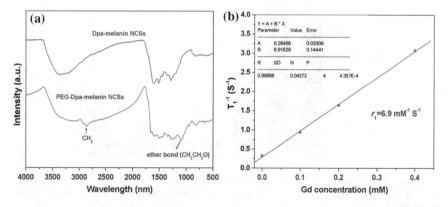

**Fig. 6.19** **a** FTIR spectra of Dpa-melanin CNSs before and after modification with bisamino-terminated PEG. **b** The linear relationship between $T1$ relaxation rates ($T1^{-1}$) and $Gd^{3+}$ ion concentrations for Gd-DTPA-modified Dpa-melanin CNSs. (Copyright 2013 Wiley-VCH)

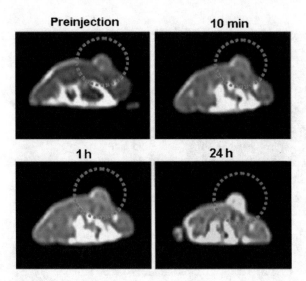

**Fig. 6.20** *In vivo* T1-weighted MR images of the 4T1-tumor bearing mouse before and after intravenous injection of the Gd-DTPA-modified Dpa-melanin CNSs solution. The red circles point the tumor sites. (Copyright 2013 Wiley-VCH)

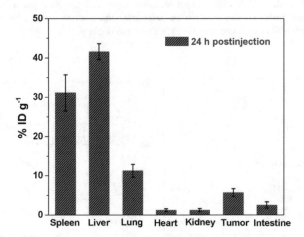

**Fig. 6.21** Biodistribution of Gd-DTPA-modified Dpa-melanin CNSs in 4T1 tumor-bearing mice at 24 h after intravenous injection of Gd-DTPA-modified Dpa-melanin CNSs. (Copyright 2013 Wiley-VCH)

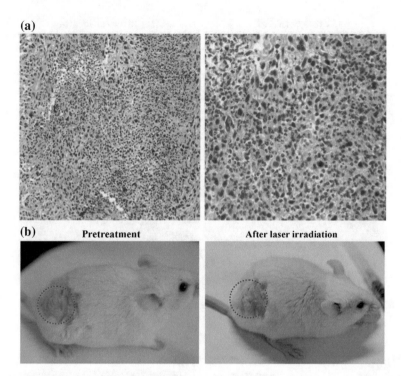

**Fig. 6.22 a** Representative H&E stained histological images of the tumor section at different magnifications from the mouse after intravenous injection of Gd-DTPA-modified Dpa-melanin CNSs and laser irradiation at 24 h postinjection. **b** Digital photos of a 4T1 tumor-bearing mouse that received intravenous injection of Gd-DTPA-modified Dpa-melanin CNSs before and after photothermal therapy at 24 h postinjection. (Copyright 2013 Wiley-VCH)

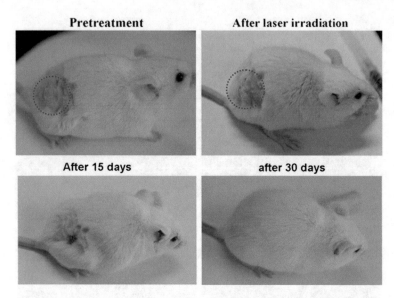

**Fig. 6.23** Digital pictures of 4T1 tumor-bearing mouse after photothermal treatment

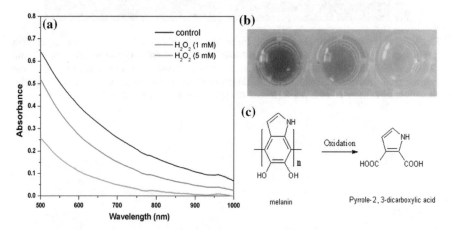

**Fig. 6.24** **a** UV-vis absorption spectra of Dpa-melanin CNSs dispersion before and 24 h after addition of $H_2O_2$. **b** The corresponding photograph of Dpa-melanin CNSs from a. **c** Proposed chemical degradation pathway of Dpa-melanin CNSs by $H_2O_2$. (Copyright 2013 Wiley-VCH)

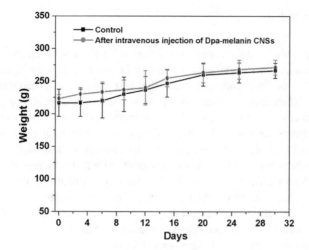

**Fig. 6.25** Body weight changes of the rats with versus without intravenous injection of Dpa-melanin CNSs versus time. (Copyright 2013 Wiley-VCH)

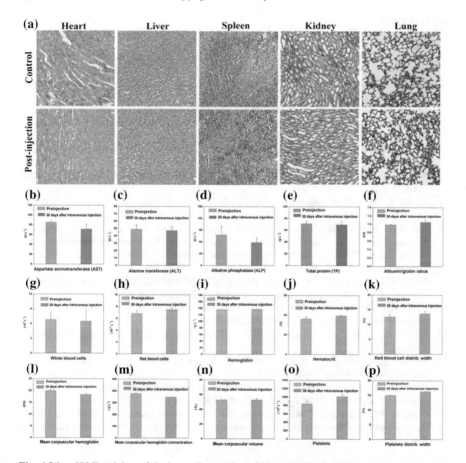

**Fig. 6.26 a** H&E staining of the heart, liver, spleen, kidney, and lung of the rat one month after intravenous injection of a single dose of Dpa-melanin CNSs solution. **b–p** Blood test parameters for control rats and Dpa-melanin CNS-treated groups. (Copyright 2013 Wiley-VCH)

## 6.4    Conclusion

In this chapter, we have described an innovative PTT agent based on Dpa-melanin CNSs. Differing from previously studied PTT agents, Dpa-melanin CNSs are composed of naturally occurring melanin that was widely distributed throughout the human body. Thus, it could effectively avoid serious adverse effects associated with the long-term retention of foreign substances in the body. Dpa-melanin CNSs could be easily prepared and well dispersed in aqueous media with high colloidal stability. Furthermore, they showed strong NIR absorption and high photothermal conversion efficiency of 40%. They efficiently killed the cancer cells and suppressed the tumor growth without obvious toxicity. In addition, they possessed a high LD50 value and were biodegradable. After modification with Gd-DTPA, simultaneous MRI and photothermal therapy of tumor were achieved. We expected that Dpa-melanin CNSs can provide an effective, powerful tool for imaging-guided therapy of advanced human diseases including cancers.

## References

1. Yoo D, Lee JH, Shin TH et al Theranostic magnetic nanoparticles. Acc Chem Res 44: 863–874
2. Vogel A, Venugopalan V (2003) Mechanisms of pulsed laser ablation of biological tissues. Chem Rev 103:577–664
3. Nolsøe CP, Torp-Pedersen S, Burcharth F et al (1993) Interstitial hyperthermia of colorectal liver metastases with a US-guided Nd-YAG laser with a diffuser tip: a pilot clinical study. Radiology 187:333–337
4. Lal S, Clare SE, Halas NJ (2008) Nanoshell-enabled photothermal cancer therapy: impending clinical impact. Acc Chem Res 41:1842–1851
5. Nel A, Xia T, Madler L et al (2006) Toxic potential of materials at the nanolevel. Science 311:622–627
6. Sharifi S, Behzadi S, Laurent S et al (2012) Toxicity of nanomaterials. Chem Soc Rev 41:2323–2343
7. Lovell JF, Jin CS, Huynh E et al (2011) Porphysome nanovesicles generated by porphyrin bilayers for use as multimodal biophotonic contrast agents. Nat Mater 10:324–332
8. Simon JD (2000) Spectroscopic and dynamic studies of the epidermal chromophores trans-urocanic acid and eumelanin. Acc Chem Res 33:307–313
9. Liu Y, Simon JD (2003) Isolation and biophysical studies of natural eumelanins: applications of imaging technologies and ultrafast spectroscopy. Pigment Cell Res 16:606–618
10. Chen HJ, Shao L, Ming TA et al (2010) Understanding the photothermal conversion efficiency of gold nanocrystals. Small 6:2272–2280
11. Blois MS, Maling JE, Zahlan AB (1964) Electron spin resonance studies on melanin. Biophys J 4:471–490
12. Fisher OZ, Larson BL, Hill PS et al (2012) Melanin-like hydrogels derived from gallic macromers. Adv Mater 24:3032–3036
13. Peter MC, Forsfer H (1989) On the structure of eumelanins: identification of constitutional patterns by solid-state NMR spectroscopy. Angew Chem Int Ed 28:741–743
14. Centeno SA, Shamir J (2008) Surface enhanced Raman scattering (SERS) and FTIR characterization of the sepia melanin pigment used in works of art. J Mol Struct 873:149–159

15. Zhao Y, Pan H, Lou Y et al (2009) Plasmonic $Cu_{2-x}S$ nanocrystals: optical and structural properties of copper-deficient copper(I) sulfides. J Am Chem Soc 131:4253–4261
16. Habash RWY, Bansal R, Krewski D et al (2006) Thermal therapy, part 1: an introduction to thermal therapy. Crit Rev Biomed Eng 34:459–489
17. Roper DK, Ahn W, Hoepfner M (2007) Microscale heat transfer transduced by surface plasmon resonant Gold nanoparticles. J Phys Chem C 111:3636–3641
18. Cave AC, Brewer AC, Narayanapanicker A et al (2006) NADPH oxidases in cardiovascular health and disease. Antioxid Redox Signal 8:691–728

Printed in the United States
By Bookmasters